Lecture Notes in Computer Science   8060

Commenced Publication in 1973
Founding and Former Series Editors:
Gerhard Goos, Juris Hartmanis, and Jan van Leeuwen

T0202505

Miroslav Bursa   Sami Khuri
M. Elena Renda (Eds.)

Information Technology in Bio- and Medical Informatics

4th International Conference, ITBAM 2013
Prague, Czech Republic, August 28, 2013
Proceedings

 Springer

Volume Editors

Miroslav Bursa
Czech Technical University in Prague
Faculty of Electrical Engineering
Department of Cybernetics
Technicka 2, 166 27 Prague 6, Czech Republic
E-mail: bursam@fel.cvut.cz

Sami Khuri
San Jose State University
Department of Computer Science
One Washington Square, San Jose, CA 95192-0249, USA
E-mail: khuri@cs.sjsu.edu

M. Elena Renda
Istituto di Informatica e Telematica del CNR
Via G. Moruzzi 1, 56124 Pisa, Italy
E-mail: elena.renda@iit.cnr.it

ISSN 0302-9743   e-ISSN 1611-3349
ISBN 978-3-642-40092-6   e-ISBN 978-3-642-40093-3
DOI 10.1007/978-3-642-40093-3
Springer Heidelberg Dordrecht London New York

Library of Congress Control Number: 2013944541

CR Subject Classification (1998): J.3, H.2, H.3, H.4, I.2

LNCS Sublibrary: SL 3 – Information Systems and Application,
incl. Internet/Web and HCI

Typesetting: Camera-ready by author, data conversion by Scientific Publishing Services, Chennai, India

Printed on acid-free paper

Springer is part of Springer Science+Business Media (www.springer.com)

Preface

Biomedical engineering and medical informatics represent challenging and rapidly growing areas. Applications of information technology in these areas are of paramount importance. Building on the success of ITBAM 2010, ITBAM 2011, and ITBAM 2012, the aim of the fourth ITBAM conference was to continue bringing together scientists, researchers, and practitioners from different disciplines, namely, from mathematics, computer science, bioinformatics, biomedical engineering, medicine, biology, and different fields of life sciences, so they can present and discuss their research results in bioinformatics and medical informatics. We believe that ITBAM 2013 served as a platform for fruitful discussions between all attendees, where participants could exchange their recent results, identify future directions and challenges, initiate possible collaborative research and develop common languages for solving problems in the realm of biomedical engineering, bioinformatics, and medical informatics. The importance of computer-aided diagnosis and therapy continues to draw attention worldwide and has laid the foundations for modern medicine with excellent potential for promising applications in a variety of fields, such as telemedicine, Web-based healthcare, analysis of genetic information, and personalized medicine.

Following a thorough peer-review process, we selected seven long papers for oral presentation and six short papers for the poster session for the fourth annual ITBAM conference. The Organizing Committee would like to thank the reviewers for their excellent job. The articles can be found in the proceedings and are divided into the following sections: Critical Health and Intelligent Systems in Medical Research and Obstetrics, Neonatology and Decision Systems in Cardiology. The papers show how broad the spectrum of topics in applications of information technology to biomedical engineering and medical informatics is.

The editors would like to thank all the participants for their high-quality contributions and Springer for publishing the proceedings of this conference. Once again, our special thanks go to Gabriela Wagner for her hard work on various aspects of this event.

June 2013

Miroslav Bursa
Sami Khuri
M. Elena Renda

Organization

Program Chairs

Miroslav Bursa   Czech Technical University Prague,
                                     Czech Republic
Sami Khuri   San José State University, USA
M. Elena Renda   IIT - CNR, Pisa, Italy

General Chair

Christian Böhm   University of Munich, Germany

Program Committee

Werner Aigner	FAW, Austria
Tatsuya Akutsu	Kyoto University, Japan
Peter Baumann	Jacobs University Bremen, Germany
Veselka Boeva	Technical University of Plovdiv, Bulgaria
Gianluca Bontempi	Université Libre de Bruxelles, Belgium
Roberta Bosotti	Nerviano Medical Science s.r.l., Italy
Christian Böhm	University of Munich, Germany
Rita Casadio	University of Bologna, Italy
Sònia Casillas	Universitat Autònoma de Barcelona, Spain
Kun-Mao Chao	National Taiwan University, Taiwan
Vaclav Chudacek	Czech Technical University in Prague, Czech Republic
Hans-Dieter Ehrich	Technical University of Braunschweig, Germany
Maria Federico	University of Modena and Reggio Emilia, Italy
Christoph M. Friedrich	University of Applied Sciences Dortmund, Germany
Alejandro Giorgetti	University of Verona, Italy
Volker Heun	Ludwig-Maximilians-Universität München, Germany
Lars Kaderali	University of Technology Dresden, Germany
Alastair Kerr	University of Edinburgh, UK
Michal Krátký	Technical University of Ostrava, Czech Republic
Vaclav Kremen	Czech Technical University in Prague, Czech Republic

Jakub Kuzilek   Czech Technical University, Czech Republic
Gorka Lasso   CICbioGUNE, Spain
Lenka Lhotska   Czech Technical University, Czech Republic
Roger Marshall   Plymouth State University, USA
Elio Masciari   ICAR-CNR, Università della Calabria, Italy
Erika Melissari   University of Pisa, Italy
Henning Mersch   RWTH Aachen University, Germany
Aleksandar Milosavljevic   Baylor College of Medicine, USA
Jean-Christophe Nebel   Kingston University, UK
Vit Novacek   National University of Ireland, Galway, Ireland
Nadia Pisanti   University of Pisa, Italy
Cinzia Pizzi   Università degli Studi di Padova, Italy
Clara Pizzuti   Institute for High Performance Computing and
                                 Networking (ICAR)-National Research
                                 Council (CNR), Italy

Nicole Radde   Universität Stuttgart, Germany
Roberto Santana   University of the Basque Country (UPV/EHU),
                                 Spain
Kristan Schneider   University of Vienna, Austria
Huseyin Seker   De Montfort University, UK
Jiri Spilka   Czech Technical University in Prague,
                                 Czech Republic
Kathleen Steinhofel   King's College London, UK
Karla Stepanova   Czech Technical University, Czech Republic
Viacheslav Wolfengagen   Institute JurInfoR-MSU, Russian Federation
Borys Wrobel   Polish Academy of Sciences, Poland
Filip Zavoral   Charles University in Prague, Czech Republic
Songmao Zhang   Chinese Academy of Sciences, China
Qiang Zhu   The University of Michigan, USA
Frank Gerrit Zoellner   University of Heidelberg, Germany

Table of Contents

Poster Session

Pervasive and Intelligent Decision Support in Critical Health Care Using Ensembles

Filipe Portela[1], Manuel Filipe Santos[1], José Machado[2],
António Abelha[2], and Álvaro Silva[3]

[1] Algoritmi Centre, University of Minho, Portugal
{cfp,mfs}@dsi.uminho.pt
[2] CCTC, University of Minho, Portugal
{jmac,abelha}@di.uminho.pt
[3] Serviço Cuidados Intensivos, Centro Hospitalar do Porto, Hospital Santo António
moreirasilva@clix.pt

Abstract. Critical health care is one of the most difficult areas to make decisions. Every day new situations appear and doctors need to decide very quickly. Moreover, it is difficult to have an exact perception of the patient situation and a precise prediction on the future condition. The introduction of Intelligent Decision Support Systems (IDSS) in this area can help the doctors in the decision making process, giving them an important support based in new knowledge. Previous work has demonstrated that is possible to use data mining models to predict future situations of patients. Even so, two other problems arise: i) how fast; and ii) how accurate? To answer these questions, an ensemble strategy was experimented in the context of INTCare system, a pervasive IDSS to automatically predict the organ failure and the outcome of the patients throughout next 24 hours. This paper presents the results obtained combining real-time data processing with ensemble approach in the intensive care unit of the Centro Hospitalar do Porto, Porto, Portugal.

1   Introduction

The use of Data Mining (DM) embebed in intelligent systems is no more an innovation. This concept is widely explored and consolidated. However, the idea of developing Intelligent Decision Support Systems (IDSS) by combining online-learning, DM ensambles, real-time processing and pervasiveness in a critical area, such is Intensive Medicine, is a huge challenge. This approach can represent an important step to provide the best decision in the interest of the pacient.

The previous work carried out by this research team and the promising results attained in this constitute the principal motivation for this new work: deciding in real-time using online data. In this context was developed the INTCare system, a pervasive IDSS for Intensive Care Medicine (ICU) that can, automatically and in real-time, predict patient organ failure (renal, hepatic, neurologic, respiratory, cardiovascular, coagulation) and patient outcome for the next 24 hours [1, 2]. This system makes use of a set of intelligent agents that are responsible for performing autonomous tasks such as, for example, data acquisition and the entire knowledge discovery process. In order to improve the accuracy associated to the predictive models, a promising

M. Bursa, S. Khuri, and M.E. Renda (Eds.): ITBAM 2013, LNCS 8060, pp. 1–16, 2013.

approach called ensemble learning [3-6] was explored. The ensemble is used to ensure that the system is always using the best DM results. All the results obtained by the DM models are analyzed and the best solution for each target is selected. In this work 7 models were induced using 3 different techniques (Decision Trees, Support Vector Machine and Naive Byes) for each one of the six targets (outcome, renal, hepatic, respiratory, coagulation and cardiovascular). As a way to evaluate the results, three metrics were considered: sensitivity, accuracy and total errors. These measures were defined in accordance with ICU doctors. Sensitivity is considered the most important measure to select models because in critical health care doctors want to know with a high precision how bad the patient can be. The ensemble helps choosing the best model for each target, i.e., the model with the highest sensitivity. The information obtained is disseminated through a situated display installed in the ICU Decision Room, where the user can consult the information. This system is currently implemented in a real environment at the ICU of CHP in the city of Porto, Portugal.

Next sections depict in detail how ensemble has been implemented. The results obtained during the experimentation are discussed and compared with those obtained by the monolithic approach [2, 7].

Section 2 presents previous work made by the research team and some relevant concepts. In section 3 a characterization of the data used to induce the predictive models is object of study. Section 4 presents the ensemble setting up. Section 5 presents experimental work performed in the ICU. The results obtained so far are discussed in the subsequent section. Finally, section 7 concludes the work and points some future work.

2   Background

2.1   Intelligent Decision Support in Intensive Care Medicine

The main objective of intensive care medicine is to avoid / reverse the organ failure and retrieving the patient to the best possible conditions [8]. Usually, this type of medicine is applied in Intensive Care Units (ICU). The ICU environment is categorized as critical due to the need of deciding / acting quickly and carefully facing complex situations [9]. The professionals of the ICU (nurses and doctors) claim for a system that minimizes the manual efforts and don't interfere in the daily patient care tasks [10, 11]. Patient direct care is the first concern, while the data documentation is in a second landing [12]. The introduction of an IDSS in this type of environments should to be done too carefully. An IDSS not synchronized with the activities performed in ICU can compromise the quality of the decision. Nowadays, a large number of data are still registered in paper preventing the implementation of pervasive IDSS. Excepting the INTCare system, there is no other system with similar characteristics working in an ICU real environment.

2.2   INTCare

INTCare is a research project started in 2008 and has been responsible for significant changes in the ICU environment (e.g. online-learning, automatic and real-time data acquisition and data processing, pervasiveness and others) [13]. The main goal of INTCare is to predict the failure of six organic systems (cardiovascular, coagulation, renal, hepatic, neurologic and respiratory) and the outcome of the patients along the next hour. This goal was achieved making use of artificial intelligence techniques like: intelligent agents and data mining. The prior results were obtained with an offline learning approach using the EURICUS database [14].

Further developments included an automatic acquisition of data [15] [2] and online-learning. Nowadays, the predictive models are able to make the predictions along the next 24 hours, i.e. the doctors can foresee for the next 24 hours, in an hourly base mode, the probability associated to the organ failure/dysfunction and outcome of patients.

Fig 1 presents the overall architecture of the INTCare system in terms of modules and agents. Basically, the system receives the data collected by the agents from data acquisition sub-system and furnishes predictions based on this data. These predictions are furnished using the agents present in the knowledge and inference sub-systems. Finally, the results (in terms of probabilities) are presented in the INTCare interface. This system is being used in the ICU of CHP.

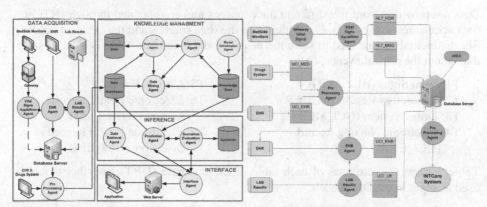

Fig. 1. INTCare System   **Fig. 2.** Data Acquisition process

2.3   Data Mining Techniques and Tool

In this project it was used the Oracle Data Mining (ODM) tool [16, 17]. ODM can meet all DM tasks using a convenient connection with the database. The data mining engine receives the data stored into the database tables, processes the models automatically and inserts the probability / prediction results in the DM Table. From the available DM techniques, only three were considered for this project:

- Decision Trees (DT) - C4.5 algorithm [18];
- Support Vector Machine (SVM) - Vapnik-Chervonenkis algorithm [19];
- Naive Byes (NB) [17].

To assess and compare the models confusion matrix and Receiver Operating Characteristic (ROC) curve were considered [20].

2.4 Ensembles

The ensemble-learning methodology consists in two sequential phases: the training and the testing phase [3]. During the training phase, several different predictive models are generated from the training set. In the test phase, the ensemble is executed and aggregates the outputs for each predictive model [3]. In this project the ensemble compares the models and defines which one is the best for each target. The different types of models can contribute to decrease the number of coincident failures and increasing the ensemble performance [21]. In this project was followed the Stacked Generalization methodologies [22, 23]. The learning procedure is composed by four steps [3]: 1) Split the training set into two disjoint sets; 2) Train several base learners on the 1st part; 3) Test the base learners on the 2nd part; 4) Using the predictions made in the stage 3 as the inputs and the correct responses as the outputs, train a higher level learner. Stratification technique was used to split the dataset. For each target a different dataset was considered with the same distribution as the target classes (0, 1).

3 Data Characterization

Data are provided from five different data sources and are obtained through INTCare data acquisition agents [24]. Fig. 2 identifies the data sources and the agents involved. Six distinct tables are used to store the acquired data; an extra table is used to store data about the critical events (T5):

T1: Vital Signs (HL7_MSG)
T2: Laboratory Results (UCI_LR)
T3: Drugs System (UCI_MED)
T4: Electronic Nursing Record (UCI_ENR)

T5: Critical Events
T6: Electronic Health Record – Admission (UCI_EHR)
T7: Electronic Health Record – Discharge (UCI_EHR)

Table 1 maps the attributes of each table of the database (DBT) with the variables used as input in the DM models.

Table 1. Database Attributes vs. Variables of DM Models

Variable	T1	T2	T3	T4	T5	T6	T7
Patient Identification (PID) and Date	X	X	X	X	X	X	X
Blood Pressure (BP); Temperature	X			X			
Heart Rate (HR); Saturation of Oxygen (SPO2)	X			X			
Creatinine; Bilirubin; Blood Platelets		X					
Dopamine; Dobutamine; Noradrenaline			X	X			
Urine Output; Glasgow				X			
Event All (date, type, id ,...)					X		
Admission type (admT); Born date (Age)						X	
Admission from (admF)						X	
Patient Outcome (died)							X

The variables in use are presented in table 1 and are provided from five information systems: Laboratory (LR), Pharmacy System (PHS), Electronic Nursing Records (ENR), Electronic Health Records (EHR) and Vital Signs Monitor (VS).

DBT represent all of the raw values collected and stored in the database that will be used by DM engine.

$$DBT = <PID_i, Date_i, d_{BP}, d_{Temperature}, d_{HR}, d_{SPO2}, d_{UrineOutput}, d_{GSC}, d_{Dopamine}, d_{Dobutamine},$$
$$d_{Norepidnephrine}, d_{Adrenaline}, d_{EvID}, d_{EvTime}, d_{EvType}, d_{BirthdayDate}, d_{admF}, d_{admT}, d_{died} >$$

Where,

PID is the patient identification;

$Date$ is the date of the value collected;

i is the concatenation of patient id and the hour of the day;

d_{BP} ... d_{died} are the values collected for each i.

After the data are correctly collected, they are submitted to pre-processing and transforming operations. These operations are totally autonomous, made in real-time and performed by two agents: pre-processing and data mining agents (fig. 1). The tasks involved can be resumed as: 1) Validation of all data collected – only are considered the values in the normal range and having a correct PID; 2) Preparation of the data mining input table (DMIT) – the input data for each patient are created (24 rows per day are inserted); 3)Transformation of variables according to the DM attributes: CM Variables; SOFA variables; Critical Events and Ratios; 4) Validation of DM input data – clean null and wrong values to ensure that the dataset is real.

The first task consists in a data validation by the system procedures [1]. This procedure verifies if the values collected are within ICU ranges and if they have assigned a correct patient id. In the second task DMIT is created (120 rows by patient and one column for each DM variable). The third task takes into account the variables used by DM and transforms and inserts the data collected into the DMIT. Derived variables / values obtained correspond to:

- ✓ **Case Mix (CM)** {Age, Admission type, Admission from};
- ✓ **SOFA** {Cardio, Respiratory, Renal, Liver, Coagulation, neurologic}
- ✓ **Critical Events Accumulated (ACE)** {ACE of Blood Pressure (BP), ACE of Oxygen Saturation (SO2), ACE of Heart Rate (HR) and ACE of Urine Output (UR), Total ACE};
- ✓ **Ratios1 (R1)** {ACE of BP/elapsed time of stay, ACE of SO2/elapsed time of stay , ACE of HR/elapsed time of stay, ACE of UR/elapsed time of stay, Total of ACE / elapsed time of stay};
- ✓ **Ratios2 (R2)** {ACE of BP / max number of ACE of BP , ACE of SO2/ max number of ACE of SO2 , ACE of HR / max number of ACE of HR, ACE of Ur / max number of UR, Total ACE/ max number of total ACE }
- ✓ **Ratios (R)** is a union of the two ratios set: **R** = R1 U R2.
- ✓ **Outcome** {died};

Where

- ✓ ACE is the number of accumulated Critical Events per patient during their admission by hour and type;
- ✓ elapsed time of stay is the number of hours since patient admission until now;
- ✓ max number of ACE is the maximum number of ACE verified by a patient per hour. These values are refreshed anytime a worse value appears.

In task 3 the values of the variables are transformed into discrete and normalized values in order to be used by different DM techniques. Table 3 presents for each one the variables and the list of values to be considered according to a range of values (min, max). For the CM attributes, it is only used one value for each case. For example, if a patient is 48 years old and has an urgent admission from emergency department, the transformed values are 2, u, 3 respectively for each ID column. For the real numbers (values $\in \{|R0+\}$), some conjuncts were introduced according to the normality and importance /significance of the value to the ICU. For ACE, R1 and R2 the transformed value will be calculated according to the min and max limits defined in the table III. For example, in the case of the renal system if the creatinine value for a given patient is less or equal than the Min (see table 2), the value of RenalSofa variable is 1 otherwise is 0. For the accumulated events (ACE), it is used the total number of the events verified before the admission time. Finally, in the case of the outcome the default is 0, except when the patient dies (value 1).

Table 2. DMT Attributes, Frequency and data source of the variables considered

	ID	Source	Frequency	Variable	Min	Max	Value
CASE MIX	Age	EHR	Once	-	18	46	1
				-	47	65	2
				-	66	75	3
				-	76	130	4
	Admin Type	EHR	Once	Urgent	-	-	u
				Programed	-	-	p
				Chirurgic	-	-	1
				Observation	-	-	2
	Admin From	EHR	Once	Emergency	-	-	3
				Other ICU	-	-	4
				Other Hospital	-	-	5
				Other Situation	-	-	6
SOFA	Cardio	VS	Minute	BP (mean)	0	70	1
		PHS	All day	Dopamine	0,01	-	1
		PHS	All day	Dobutamine	0,01	-	1
		PHS	All day	Epi / Norepi	0,01	-	1
	Renal	LR	All day	Creatinine	1.2	-	1
	Resp	LR	All day	Po2/Fio2	0	400	1
	Hepatic	LR	All day	Bilirubin	1.2	-	1
	Coagul	LR	All day	Platelets	0	150	1
	Neuro	ENR	All day	Glasgow	3	14	1
	ACE	VS / ENR	Hour / Minute	-	0	+oo	SET
	R1	VS / ENR	All day	-	0	1	SET
	R2	VS / ENR	All day	-	0	1	SET
	Outcome	EHR	All day	Live or Dead	-	-	0 or 1

Table 3 presents the discretization rules defined for each continuous value (values ∈ {|R0+}). The ranges were created using a 7-point-scale adapted by Clinical Global Impression - Severity scale (CGI-S) [25]. The goal of CGI-S is to allow clinician for rating the severity of illness [26]. The boundaries of each set were defined after some meetings with ICU doctors and after analyzing the significance of each set of values. More severe cases are assigned to the levels 6 and 7. At the top of the table is the identification of the set. The left column identifies the variable. In the middle of the table the ranges for each set are defined.

R1min and R1max are used by R1 (max number of ACE). It is categorized according to the percentage of the collected value. For example, for the R1 attribute 'ACE of BP / max number of ACE of BP', if a patient has 7 ACE at the sixth hour and the maximum verified in the past for this time is 10 ACE, the respective set, according to table IV, will be 4 (7/10 = 0,7). For all attributes of R1, the ranges of the set are equal. For R2 attributes (ratios that use the elapse time), it is used the rows (R2 BP min to R2 TOT max) to determine the set. In this case, each attribute has a different range. For the attribute 'ACE of O2/elapsed time of stay' if, for example, a patient has 10 ACE of O2 at the 50th hour, the ratio value will be 0,2 and the DM set will be 5 (0,1<0.2<=0,3).

Finally, all ACE values are grouped in accordance to their importance and number. For example, a patient that has 8 ACE, using table 3 and ace row, corresponds to the set 3 (5 < 8 <=8) in the DM input table.

Table 3. Discretization set of Data Mining Inputs

SET		0	1	2	3	4	5	6	7
R1	Min	-0,1	0	0,2	0,4	0,6	0,8	1	-
	Max	0	0,2	0,4	0,6	0,8	1	+00	-
R2	Min	-0,1	0,000	0,020	0,040	0,075	0,100	0,300	0,500
BP	Max	0,000	0,020	0,040	0,075	0,100	0,300	0,500	1
R2	Min	-0,1	0,000	0,020	0,040	0,075	0,100	0,300	0,500
O2	Max	0,000	0,020	0,040	0,075	0,100	0,300	0,500	1
R2	Min	-0,1	0,000	0,001	0,003	0,006	0,010	0,030	0,100
HR	Max	0,000	0,001	0,003	0,006	0,010	0,030	0,100	1
R2	Min	-0,1	0,000	0,020	0,050	0,080	0,100	0,300	0,500
UR	Max	0,000	0,020	0,050	0,080	0,100	0,300	0,500	1
R2	Min	-0,1	0,000	0,020	0,050	0,080	0,100	0,300	0,300
TOT	Max	0,000	0,020	0,050	0,080	0,100	0,300	0,500	1
ACE	Min	-0,1	0	3	5	8	10	12	15
	Max	0	3	5	8	10	12	15	50

4   Ensemble Design

This chapter introduces the ensemble considered, how it is structured and the way how it works.

4.1   Predictive Models

Making use of the four groups of attributes associated to *DMIT*, 126 models were developed. All models contain the PID, Date, and the hours since the patient admission into the ICU until the discharge during a maximum of 5 days (0-120). The variables considered for each model are the following:

MI =	CM U ACE	*MV* =	CM U ACE U SOFA U R
MII =	CM U ACE U R	*MVI* =	CM U ACE U SOFA U R2
MIII =	CM U ACE U R1	*MVII* =	CM U ACE U SOFA U R1
MIV =	CM U ACE U SOFA		

Each model is obtained using the variables presented in the DMIT and is represented by the tuple:

$$DMIT = < pid, date, hour, v_{ace\_bp}, v_{aceBP\_time}, v_{aceBP\_max}, v_{ace\_hr}, v_{acehr\_time}, v_{acehr\_max},$$
$$v_{ace\_spo2}, v_{acespo2\_time}, v_{acespo2\_max}, v_{ace\_ur}, v_{aceur\_time}, v_{aceur\_max}, v_{total\_ace}, v_{totalace\_time},$$
$$v_{totalace\_max}, v_{age}, v_{adminF}, v_{adminT}, v_{sofa\_cardio}, v_{sofa\_resp}, v_{sofa\_renal}, v_{sofa\_coag}, v_{sofa\_hepa},$$
$$v_{sofa\_neuro}, v_{outcome} >$$

Where,
    pid is the patient identification;
    date is the date of the values;
    hour is the number of hours elapsed since the patient admission;
    $v_{ace\_bp} ... v_{outcome}$ are the values obtained for each patient and date.

The next representation is an example of M7 using decision trees (DT) for the respiratory system:

$$DMIT = < pid, date, hour, v_{age}, v_{adminF}, v_{adminT}, v_{sofa\_cardio}, v_{sofa\_renal}, v_{sofa\_coag}, v_{sofa\_hepa},$$
$$v_{sofa\_neuro}, v_{ace\_bp}, v_{aceBP\_time}, v_{aceBP\_max}, v_{ace\_hr}, v_{acehr\_time}, v_{acehr\_max}, v_{ace\_spo2}, v_{acespo2\_time},$$
$$v_{acespo2\_max}, v_{ace\_ur}, v_{aceur\_time}, v_{aceur\_max} >$$

After the DMIT is filled and validated, the models are induced online by the DM agent. This agent runs whenever a request is sent or when it verifies that the performance of the models is decreasing.

4.2   Ensemble

Ensemble can be defined as a three-dimensional matrix *Mx* composed by m=7 models (MI to MVII) x t=6 targets (tI to tVI) x z=3 (zI to zIII) techniques. Each cellule characterizes a particular model by:

$$M_{m,t,z} = \begin{cases} m = I ... VII \\ t = I ... VI \\ z = I ... III \end{cases} \quad Ensemble_{mtz} = \begin{bmatrix} model(mtz), accuracy(mtz), sensibility(mtz), \\ terror(mtz), specificity(mtz) \end{bmatrix}$$

Fig. 3 presents an overview of how the ensemble works. The process is divided in two major phases:

✓ Predictive models – the predictive models MI...MVII are trained making use of three different techniques: SVM, DT and NB. This task generates 7x3x6=126 models.

✓ Ensemble – the models are assessed in terms of the sensitivity, accuracy, total error and specificity. The best model for each target is then selected. A different ensemble is defined for each target.

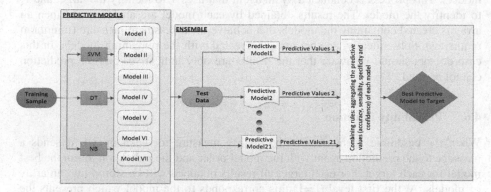

Fig. 3. Ensemble main processing cycle structure and process

The tasks performed in the main cycle are: 1) Create a Confusion matrix for each one of the models; 2) Obtain the assessment measures; 3) Evaluate the models and compare the results; 4) Define the probability function for the best model of each target.

4.3   Confusion Matrix

At the first step the confusion matrix function of ODM is automatically used. In this way the following results are obtained: the number of true positive (TP), false positive (FP), true negatives (TN) and false negatives (FN) presented in the dataset. These results are obtained for each model (k).

4.4   Assessment Measures

In this task a procedure calculates a set of measures using the TP, FP, FN and FP values. The measures calculated are: sensitivity, specificity, accuracy, and predictive confidence. All the measures has the same level of importance for all models (k):

$$SENSITIVITY(K) = TP(K) / (TP(K) + FN(K))$$
$$ACCURACY(K) = (TP(K) + TN(K)) / (TP(K) + TN(K) + FN(K) + FP(K))$$
$$TERROR(K) = FP(K) + FN(K) / (TP(K) + TN(K) + FN(K) + FP(K))$$

4.5   Evaluating Models and Compare the Results

The main purpose of the ensemble is to select the best model from a set of models. In order to achieve this goal a set of thresholds are defined for each measure, according to ICU doctors: Sensitivity $>= 0.85$; Accuracy $>= 0.75$ and Terror $<= 0.40$.

The objective is to obtain models most sensible as possible, having a minimum boundary of accuracy ($>=0,75$) and providing a low level of errors.

Ensemble is executed for each target (tI..tVII). Models are distinguished into two sets: one group includes the models that achieve the thresholds; and the remaining models. This process is conducted by an agent that uses T to identify the target and K to identify the model. The results obtained by each model are analysed and then an array is created containing the models that achieve the assessment measure (minimum requisites). Subsequently, another array is created with the remaining models. In this case doctors should be aware that the results are only indicatives and the prediction cannot be used.

4.6   Probability Function

When the ensemble agent obtains the quality measure for each model, it sends a message to the prediction agent with the model order and the indication to run the best model for each target. For each target all results obtained will be stored into an array of models. At the first level ($L=1$) this corresponds to the model which presents the best result. As the result it is generated a view for each target (t).

$DM_{target(t)} = <$ pid, date, hour, V_{ace_bp}, V_{aceBP_time}, V_{aceBP_max}, V_{ace_hr}, V_{acehr_time}, V_{acehr_max}, V_{ace_spo2}, $V_{acespo2\_time}$, $V_{acespo2\_max}$, V_{ace_ur}, V_{aceur_time}, V_{aceur_max}, V_{total_ace}, $V_{totalace\_time}$, $V_{totalace\_max}$, V_{age}, V_{adminF}, V_{adminT}, V_{sofa_cardio}, V_{sofa_resp}, V_{sofa_renal}, V_{sofa_coag}, V_{sofa_hepa}, V_{sofa_neuro}, $V_{outcome}$, $P_{respiratory}$, P_{cardio}, P_{renal}, $P_{hepatic}$, $P_{coagulation}$, $P_{outcome} >$

Where,

z is the identifications of the results ;
pid is the patient identification;
date is the date of prediction;
hour is the respective hour of the probability results
$P_{respiratory}$ ... , $P_{outcome}$ are the probability(percentage) of the target be 1 by patient and hour.

Latterly, prediction probability for each target / hour is disseminated through INTCare system. If none of the models correspond to the quality objectives for some target, prediction is not presented. When all models have their measures calculated, ensemble agent compares the models. The best model for each target should present a total error value less than 0.40, an accuracy value higher than 0.75 and sensitivity value higher than 0.85. Due to the dynamic characteristics of the ICU environment there are a high number of false positives and false negatives. The patient condition changes over the time and sometimes these changes are influenced by the clinical proceedings performed by nurses and doctors. The effects of these procedures can result in patient recovery and consequently contribute for FP or FN. However, is a medical imposition that models can't have a total percentage of error higher than

40%. The sensitivity can't be lower than 85% because the objective is not to give a correct answer but to predict what will happen to the patient in the next hour. Not disregarding the accuracy of the models that need to be higher than 75%. This value avoids bad predictions i.e. false positives. This value can contribute to increase the significance of the sensitivity. To compare the models, the following metrics were used (by order of importance): Sensitivity, Accuracy, Total Error and Specificity.

5 Experimental Results

A set of experiments have been carried out in order to test the ensemble performance. Online and real-time data collected from the ICU has been considered in the following conditions:

- Collection Time: 98 days;
- Number of Patients: 94;
- Time Frame Considered: first five days for each patient;
- Exclusion Criteria: from these data were excluded patients with data collecting intermittent, i.e., the collection system failed at least more than one hour in a continuous way or containing null values.

Tables 4 and 5 present the distribution of the attributes considered. For the test phase the original dataset (DMIT) was divided into two data sets using the holdout sampling method: 70% of the data were considered for training and 30% for testing (stratified by the target). Each target has a different dataset.

Table 4. Input Case Mix Variables distribution

Attribute	1	2	3	4	5	6	U	P
Age	15,46	45,24	7,47	31,83	-	-	-	-
AdminFrom	50,53	0	17,62	14,24	2,59	15,02	-	-
AdminType	-	-	-	-	-	-	76,89	23,11

Table 5. Input ACE and Ratios Variables distribution (%)

Attribute	0	1	2	3	4	5	6	7
ACE_BP	88,83	8,78	1,33	0,66	0,13	0,05	0,22	0,00
ACEBP_TIME	88,83	3,44	4,00	2,09	0,76	0,85	0,02	0.00
ACEBP_MAX	79,18	6,67	5,07	3,06	1,71	2,33	1,98	-
ACE_HR	84,83	12,91	1,45	0,42	0,09	0,31	0,00	0,00
ACEHR_TIME	84,83	0,08	3,21	3,62	2,46	3,99	1,36	0,46
ACEHR_MAX	84,83	5,54	5,61	1,35	1,20	1,47	0,00	0,00
ACE_SPO2	79,18	15,64	2,31	1,74	0,81	0,07	0,25	0,00
ACESPO2_TIME	79,18	4,85	4,65	4,74	2,77	3,69	0,10	0,02
ACESPO2_MAX	79,18	10,23	4,99	3,05	1,27	1,28	0,00	-
ACE_UR	100	0,00	0,00	0,00	0,00	0,00	0,00	0,00
ACEUR_TIME	100	0,00	0,00	0,00	0,00	0,00	0,00	0,00
ACEUR_MAX	100	0,00	0,00	0,00	0,00	0,00	0,00	-
TOTAL_ACE	67,90	21,85	4,53	2,07	1,41	1,17	0,23	0,83
TOTALACE_TIME	67,90	6,34	10,43	4,47	2,50	7,18	1,11	0,06
TOTALACE_MAX	84,83	5,54	5,61	1,35	1,20	1,47	0,00	-

Fig. 4 presents the distribution of the classes for each one of the targets. For example, in the input dataset 30.07% of the rows has the patient outcome value equal to 1. This value does not signify that 30.07% of the patients died, because the values are related to the rows and not to the patients, but 30.07% of the records are related to death patients. In this case, only the cardiovascular system presents a high level of cases with organ failure (cardio result = 1): 62.39%. In all of the other targets, the numbers of positive cases (final result = 1) are substantially lower.

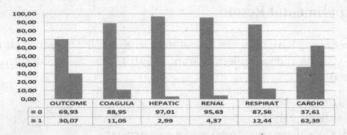

Fig. 4. Distribution of the classes

For each one of the six targets (respiratory, coagulation, renal, hepatic, cardiovascular and outcome), 21 models have been considered using three different DM techniques (SVM, DT and NB). The models have been then assessed in terms of their sensitivity, specificity, accuracy and total error. Table 6 describes the experimental settings considered for each technique. In addition this table indicates whether the used value corresponds to a default value or to a user-defined value.

Table 6. Techniques Configurations

DM Technique	Parameter Name	Parameter Value	Parameter Type
DT	Minrec Node	10	Input
	Max Depth	7	Input
	Minpct Split	.1	Input
	Impurity Metric	Gini	Input
	Minrec Split	20	Input
	Minpct Node	.05	Input
	Prep Auto	On	Input
NB	Pairwise Threshold	0	Input
	Singleton Threshold	0	Input
SVM	Conv tolerance	.001	Input
	Active learning	Al Enable	Input
	Kernel function	Linear	Default
	Complexity factor	0.142831	Default
	Prep auto	On	Input

Table 7 presents the performance achieved by the ensemble for each target. The values correspond to the average of the measures obtained during ten runs of the ensemble. To each value is associated the standard deviation. Excepting Cardiovascular system, the SVM presents the best results for sensitivity. On the other hand, M6 is the best performing model. Respiratory system doesn't meet the measures established.

Table 7. Ensemble Results

Target	Tec	Model	Sensitivity	Accuracy	Specificity	Terror
Cardio	DT	M6	**97,52±0,32**	75,04±0,65	36,77±1,57	32,85±0,83
Coagulation	SVM	M6	**88,12±1,65**	81,98±0,74	81,22±0,75	15,33±1,00
Hepatic	SVM	M3	**91,18±4,08**	83,48±0,65	83,25±0,67	12,78±2,04
Outcome	SVM	M6	**91,01±0,54**	78,24±0,58	72,74±0,87	18,13±0,44
Renal	SVM	M4	**93,43±1,85**	91,00±0,40	90,89±0,40	7,84±0,97
Respiratory	SVM	M1	**82,51±0,83**	42,30±0,61	36,60 ± 0,75	40,44±0,39

Fig. 5 presents the ROC curves obtained by target for the ten runs (average). To improve the performance of the models additional sampling techniques have been tried. The ROC curve of the ensemble is automatically obtained for each target after the creation of a table with the number and the percentage of true positive cases and false positive cases. During the test phase, some oversampling techniques were applied; however it isn't advisable due to the nature of the problem. At the same time, other grouping techniques (e.g. Bin Top 7) were used, but they presented worst results. If the ensemble doesn't find a model that meets the goals defined, the results are not presented, unless requested by the doctor. INTCare system presents, for each target, the sensitivity of the models used.

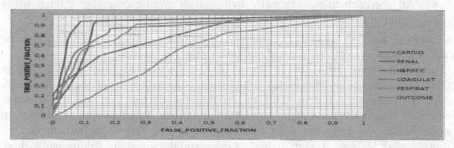

Fig. 5. ROC Curves

6   Discussion

As stated before, the models are evaluated using the following metrics (by the order of importance): Sensitivity, Accuracy, Total Error, and Specificity. In the ICU a model is considered acceptable if the following conditions are observed: Sensitivity >= 0.85; Accuracy >= 0.75 and Total Error <= 0.40. Excepting the respiratory system, all the other systems are able to be used in the ICU. In general the models present interesting values for the sensitivity. The lowest values are obtained by the respiratory and coagulation systems. The doctors can consult the results provided by ensemble engine in the INTCare system. INTCare system presents six graphs with probabilities one for each organic system and one for the outcome of patients. For each patient are presented 24 bars representing for the next 24 hours the probability associated to an organ failure/dysfunction. The patient outcome is treated and presented in the same way. Using the example of the outcome (sensitivity = 91.74), it is possible to predict that at 10 o'clock the patient will have a probability of death around 85%. In this case the doctors can give more attention to the patient and react proactively in order to recover the patient condition (avoiding death). Models with low values of specificity (e.g. cardiovascular) can generate false positive situations. This aspect is irrelevant considering the fact it detects accurately the true positive cases. The poor results observed in respiratory system are associated to the fact of the majority of the patients are with mechanical ventilation and be very difficult to automatically, understand if they are with an respiratory failure or not. The most negative aspects are related to the resources used by a patient recovering/lifesaving when the prediction is incorrect. However, this is a tolerable cost (while better models aren't available). This system is efficient for some organic system and it is very useful to help the ICU decision making process. Complementary studies on the user acceptance were carried out recurring to Technology Acceptance Model 3. The attained results were very satisfactory [27].

7   Conclusions and Future Work

This paper presented the most recent advances in the context of INTCare project. The main contributions introduced are in the form of:

- A pervasive and real-time approach for data acquisition and processing system based on agents; and
- An approach to automatically induce data mining ensemble models using online-learning.

An ensemble methodology has been explored in order to assure a good performance in the prediction of organ failure and outcome of the ICU patients.

The results were evaluated in terms of a collection of four metrics and in accordance with quality thresholds established. All targets meet these thresholds except the respiratory system. This means that the doctors can make use of the predictions to save lives and avoid non-return situations. The objective of this system is not to replace the professionals by an automated system but just supports them.

Future work includes deeper studies on the respiratory system. Additionally, more data will be considered in order to improve the models.

Acknowledgments. This work is supported by FEDER through Operational Program for Competitiveness Factors – COMPETE and by national funds though FCT – Fundação para a Ciência e Tecnologia in the scope of the project: FCOMP-01-0124-FEDER-022674. The authors would like to thank FCT (Foundation of Science and Technology, Portugal) for the financial support through the contract PTDC/EIA/72819/ 2006 (INTCare) and PTDC/EEI-SII/1302/2012 (INTCare II).. The work of Filipe Portela was supported by the grant SFRH/BD/70156/2010 from FCT.

References

1. Portela, F., Santos, M.F., Gago, P., Silva, Á., Rua, F., Abelha, A., Machado, J., Neves, J.: Enabling real-time intelligent decision support in intensive care. In: 25th European Simulation and Modelling Conference, ESM 2011, 446 p. (2011)
2. Vilas-Boas, M., Santos, M.F., Portela, F., Silva, Á., Rua, F.: Hourly prediction of organ failure and outcome in intensive care based on data mining techniques. In: ICEIS, Funchal, Madeira, Portugal (2010)

3. Kantardzic, M.: Data mining: concepts, models, methods, and algorithms. Wiley-IEEE Press (2011)
4. Gago, P., Santos, M.F.: Towards an Intelligent Decision Support System for Intensive Care Units. In: 18th European Conference on Artificial Intelligence, Greece (2008)
5. Gago, P., Santos, M.F.: Evaluating Hybrid Ensembles for Intelligent Decision Support for Intensive Care. In: Okun, O., Valentini, G. (eds.) Applications of Supervised and Unsupervised Ensemble Methods. SCI, vol. 245, pp. 251–265. Springer, Heidelberg (2009)
6. Dietterich, T.G.: Ensemble methods in machine learning. In: Kittler, J., Roli, F. (eds.) MCS 2000. LNCS, vol. 1857, pp. 1–15. Springer, Heidelberg (2000)
7. Portela, F., Pinto, F., Santos, M.F.: Data Mining Predictive Models For Pervasive Intelligent Decision Support In Intensive Care Medicine. In: KMIS 2012. INSTICC, Barcelona (2012)
8. McMillen, R.E.: End of life decisions: Nurses perceptions, feelings and experiences. Intensive and Critical Care Nursing 24, 251–259 (2008)
9. Mador, R.L., Shaw, N.T.: The impact of a Critical Care Information System (CCIS) on time spent charting and in direct patient care by staff in the ICU: A review of the literature. International Journal of Medical Informatics 78, 435–445 (2009)
10. Häyrinen, K., Saranto, K., Nykänen, P.: Definition, structure, content, use and impacts of electronic health records: A review of the research literature. International Journal of Medical Informatics 77, 291–304 (2008)
11. Orwat, C., Graefe, A., Faulwasser, T.: Towards pervasive computing in health care - A literature review. BMC Medical Informatics and Decision Making 8(26) (2008)
12. Lyerla, F., LeRouge, C., Cooke, D.A., Turpin, D., Wilson, L.: A Nursing Clinical Decision Support System and potential predictors of Head-of-Bed position for patients receiving Mechanical Ventilation. American Journal of Critical Care 19, 39–47 (2010)
13. Portela, F., Santos, M.F., Silva, Á., Machado, J., Abelha, A.: Enabling a Pervasive approach for Intelligent Decision Support in Critical Health Care. In: HCist 2011, Algarve, Portugal, p. 10 (2011)
14. Silva, Á., Cortez, P., Santos, M.F., Gomes, L., Neves, J.: Rating organ failure via adverse events using data mining in the intensive care unit. Artificial Intelligence in Medicine 43, 179–193 (2008)
15. Santos, M.F., Portela, F., Vilas-Boas, M., Machado, J., Abelha, A., Neves, J., Silva, A., Rua, F.: Information Modeling for Real-Time Decision Support in Intensive Medicine. In: Chen, S.Y., Li, Q. (eds.) Proceedings of the 8th International Conference on Applied Computer and Applied Computational Science, pp. 360–365. World Scientific and Engineering Acad. and Soć., Athens (2009)
16. Tamayo, P., Berger, C., Campos, M., Yarmus, J., Milenova, B., Mozes, A., Taft, M., Hornick, M., Krishnan, R., Thomas, S.: Oracle Data Mining. In: Data Mining and Knowledge Discovery Handbook, pp. 1315–1329 (2005)
17. Concepts, O.D.M.: 11g Release 1 (11.1). Oracle Corp. 2007 (2005)
18. Quinlan, J.R.: C4. 5: programs for machine learning. Morgan Kaufmann (1993)
19. Cherkassky, V., Mulier, F.: Vapnik-Chervonenkis(VC) learning theory and its applications. IEEE Transactions on Neural Networks 10, 985–987 (1999)
20. Bradley, A.P.: The use of the area under the ROC curve in the evaluation of machine learning algorithms. Pattern Recognition 30, 1145–1159 (1997)
21. Wang, W., Partridge, D., Etherington, J.: Hybrid ensembles and coincident-failure diversity, pp. 2376–2381. IEEE (2001)

22. Ting, K.M., Witten, I.H.: Issues in stacked generalization. Arxiv preprint arXiv:1105.5466 (2011)
23. Wolpert, D.H.: Stacked generalization*. Neural Networks 5, 241–259 (1992)
24. Portela, F., Gago, P., Santos, M.F., Silva, A., Rua, F., Machado, J., Abelha, A., Neves, J.: Knowledge Discovery for Pervasive and Real-Time Intelligent Decision Support in Intensive Care Medicine. In: Publication, S.-A.T. (ed.) KMIS 2011. Springer, Paris, France (2011)
25. Guy, W.: ECDEU assessment manual for psychopharmacology, Rockville, Md (1976)
26. Guy, W., Modified From: Rush, J., et al.: Clinical Global Impressions (CGI) Scale. Psychiatric Measures. APA (2000)
27. Portela, F., Aguiar, J., Santos, M.F., Silva, Á., Rua, F.: Pervasive Intelligent Decision Support System — Technology Acceptance in Intensive Care Units. In: Rocha, Á., Correia, A.M., Wilson, T., Stroetmann, K.A. (eds.) Advances in Information Systems and Technologies. AISC, vol. 206, pp. 279–292. Springer, Heidelberg (2013)

C-Grid: Enabling iRODS-based Grid Technology for Community Health Research

Nitin Sukhija[1] and Arun K. Datta[2,*]

[1] Mississippi State University, CAVS, Starkville, MS 39759, USA
[2] National University (NUCRI), La Jolla, CA 92037, USA
nitin@cavs.msstate.edu, adatta@nu.edu

Abstract. A Community Grid web portal, C-Grid, has been developed in this study for storing, managing and sharing large amounts of distributed community health related data in a data grid, thus facilitating further analysis of these datasets by health researchers in a collaborative environment. Remote management of this data grid is performed using the middleware iRODS, the Integrated Rule-Oriented Data System. A PHP-based wrapper, ez-iRODS, has been created as a component of C-Grid to interact with this middleware through PRODS, a client application programming interface (API). C-Grid serves as a gateway for the XSEDE resources, and also helps the users via ez-iRODS to create and manage 'virtual data collection' that can be stored in heterogeneous data resources across the distributed network. This web-based system has been developed with an objective of long-term data preservation, unified data access and sharing domain specific data amongst the scientific research collaborators of myCHOIS project.

Keywords: iRODS, cyberinfrastructure, data grid, XSEDE, community health, grid computing, data integration, portal, virtual, obesity, mobile technology.

1 Introduction

In collaboration with various Community Health Organizations, we are working on health-IT as a solution to the burgeoning problems in healthcare industry, particularly to address Childhood Obesity [1-4], which has become a national concern in US. Recent survey suggests almost 17% (or 12.5 million) of children and adolescents aged 2—19 years are obese [5]. The economic impact of this condition is staggering: in 2008 alone, it has been estimated that more than 147 billion dollars were spent just in the United States for medical costs related to obesity [6]. President Obama has recently established a task force to address this issue (White house, Office of the Press Secretary, February 09, 2010). Nevertheless, the escalating cost of providing healthcare, particularly to the patients with chronic health problems, is draining the resources of this nation. Studies suggest that the health-IT can improve the quality of healthcare if not reduce the cost by

---

* For correspondence. Preliminary work was presented at the 8th IEEE International Conference on eScience 2012, October 8 – 12, Chicago (IL), USA.

M. Bursa, S. Khuri, and M.E. Renda (Eds.): ITBAM 2013, LNCS 8060, pp. 17–31, 2013.
© Springer-Verlag Berlin Heidelberg 2013

eliminating the redundant or unnecessary services [7]. In March 2010, President Obama signed into law the Affordable Care Act that includes a wide variety of provisions designed to provide quality of healthcare. This law requires use of Electronic Health Records (EHR) for patient management by 2014 [8].

In major hospitals and professional organizations, some kind of Electronic Medical Records (EMRs) is already in use for clinical practice management. However, none of these can be claimed as EHR, as defined by the HIMSS [9]. While the data in an EMR are the legal record of what happens to a patient during their encounter at a health Care Provider Organization (CDO) and are owned by the CDO, EHR, on the other hand, is a subset of each CDO's EMR and is owned by the patient. For patients, the benefits of using EHR are obvious: convenience, portability, and efficiency [10-12]. Critical issue for implementation of EHR is data accuracy, integrity and security. In collaboration with various CDOs, National University Community Research Institute (NUCRI) is engaged in developing myCHOIS or maternity and Child Health Obesity Informatics System as a solution to these problems with a vision to offer it as a fully functional EHR in a distributed database environment.

For its School Health Program, the Department of Human Services (DHS) of the State of Illinois monitors the health of the students attending the public schools in Illinois through its schools and 59 School-based Health Clinics. In 2010 alone, health related data from about 120,000 cases have been inserted into myCHOIS database by the School Health Coordinator of DHS, which has become a valuable resource for research on childhood obesity [4]. Moreover, as a part of the preventive care, school nurses often arrange for educational training to provide valuable instructions on a variety of topics, ranging from obesity control to prevention of sexually transmitted diseases. Recently, a mobile application has been developed based on technology described earlier [3] and deployed for DHS to collect these video instructional materials.

Our partner organization, Operation Samahan, a community health organization that serves low-income families and individuals in the County of San Diego, needs IT solution for managing very large volume of health data that can be used for treating patients at the individualized level. Such medical practice, termed Individualized medicine (syn. Personalized medicine)[13], can be a reality by capturing all relevant data ranging from demography to genomics that can define a patient's health profile. If that becomes the trend, a huge amount of data will be generated on any individual patient from next generation of experiments [14]. The size of such data on any individual patient is expected to exceed terabyte to petabyte in size over his/her lifetime. As an example, each mammogram generates about 100 Mb of data and must be stored along with appropriate metadata. With the introduction of EMRs and improvements in next generation laboratory testing techniques, the quantity of personal health information that will be stored in digital form will increase dramatically [15]. The development of wearable sensors and monitoring techniques [16] will also add significantly to the volume of digital patient information.

Data storage requirement for any individual patient is predicted to be in the range of Mbyte to Tbyte per year. Given the situation where any large CDO handles thousands of patients in a year, it can easily be understandable that such enormous volumes of data can easily exceed the capacity of any local databases. Therefore, in reality, the practice will be that most of these data will be stored in a distributed fashion in a Grid [17]. The challenge will then be how those data located in heterogeneous distributed databases can be efficiently managed. The storage and computation requirements of these datasets will soon bring these CDOs to a point where traditional data management methods will become saturated. Without an efficient data management approach, users will spend most of their time just looking for data, instead of using these data to carry out their evidenced-based clinical practice. Although some advanced file systems, such as, GPFS [18], are currently available these were not designed to provide user metadata. There are several tools that have been developed to solve individual data management issues for grid computing. However, a unified environment, termed Data Grid [17][19], allows users to deal with all these issues together [20]. We have developed C-Grid to store and manage high volume of data for myCHOIS Project and deployed it for DHS and Operation Samahan.

2   Background and Related Work

The two fundamental services of a data grid are data access and metadata access. Data access services provide mechanisms to access, manage, and initiate third-party data transfer in distributed storage systems. Metadata access services provide mechanisms to access and manage information about the data stored in storage systems [21][22]. Data grid provides a user the capability of storing and managing large volume of collaborative data objects. Data or files are stored into a huge distributed network that is built using many heterogeneous storage instruments worldwide. Data located in such infrastructure are managed by a grid middleware. Among the grid middleware that are used for such data management [23-28], iRODS or Integrated Rule-Oriented Data System has become our choice. This system is designed to manage multiple file systems transparently and provides features, which help to manage user metadata [22]. Recently, a number of large scale data grid projects have been developed that use iRODS to discover, transfer, store, and manage distributed heterogeneous data. Southern California Earthquake Center [29], the Biomedical Informatics Research Network [30], GEON [31] are some of these examples. These projects utilize iRODS to transmit and manage images, videos, and a variety of data types generated in the distributed heterogeneous environments. The iRODS, acting as a grid middleware, enables collaborative data sharing and maintenance of distributed, aggregated, and integrated data storage collections.

2.1   Overview of iRODS and SRB

iRODS is an open source data grid system developed by the Data Intensive Cyber Environments (DICE) research group as a successor to Storage Resource Broker (SRB) [32] but with highly refined and newly added features. An iRODS resource is a host machine running an iRODS-server to provide data storage with rule and the microservice execution capabilities. An iRODS system includes a central iRODS Metadata Catalog or iCAT, iRODS servers, and iRODS clients.

Earlier, we designed and developed a PHP-based system, termed ezSRB, for HASS Grid Project. This project involved deployment of multiple data grids for the University of California System, an initiative of HASTAC [33]. SRB or *Storage Resource Broker* served as the middleware and was located in the data grid. For calling the functions of the SRB-server, we used S-commands to develop a PHP-wrapper. This provided a uniform client interface to different storage or file systems. The SRB also provided a metadata catalogue (MCAT) to describe and locate data within the storage systems [22]. Now, SRB has been replaced by iRODS as the middleware [32]. In contrast to SRB, iRODS is rule-oriented utilizing Rule Engine, and offers more flexibility to changing heterogeneous environments [23, 34]. Rules can be executed automatically on a defined policy enforcement time or can be invoked by users through iRODS API or rule files [35]. Therefore, a user can easily optimize the performance of an iRODS environment. It also provides management of stored data files by managing metadata for all files through an iRODS catalog, termed iCAT, which is created in a PostgreSQL server [36]. The iCAT records information about the rules, the mappings of logical file names to physical files, the metadata sets, the users and the resources managed by the iRODS servers. Also, a *perl* script is provided by iRODS, which enables replication of the metadata in the MCAT to the iCAT. In addition to the storage and metadata capabilities, iRODS provides its users with inbuilt microservices, such as, *create*, *delete*, along with rules for accessing various parts of the grid system. The iRODS users can be authenticated using the default encrypted password scheme, although the Grid Security Infrastructure [37] and Kerberos [38] can also be used. The users can interact with iRODS using the command line, iRODS Explorer (Windows only) or the iRODS Web Browser. The iRODS is written in C but there are additional programmatic interfaces available: JARGON (a Java API) [39], PRODS (a PHP API) [40] and a recent addition of PyRODS (a Python API) [41]. More about iRODS and its implementation can be viewed at its web site [42].

2.2   Overview of myCHOIS

Earlier, we developed CHOIS [1] in collaboration with the Illinois Department of Public Health (IDPH), University of Illinois at Urbana-Champaign (UIUC), the National Center for Supercomputing Application (NCSA), and the National University Community Research Institute (NUCRI), as a health IT-based solution for digitally collecting the demographic data including height and weight of school children for the Obesity Surveillance Program of the State of Illinois. Automatic computation of BMI, BMI percentile and the risk of obesity alert are embedded into this HIPAA [43] and FERPA [44] compliant secure web-based system. This system enables school-nurse and healthcare service providers to collect health records on K1-12 children and report statistical and surveillance information on BMI to identify those at-risk, overweight, or obese, and help health administrators to design and implement obesity prevention and intervention programs. Further development of this system includes fields for collecting relevant data on maternal health. This improved system is now termed myCHOIS. Earlier, this web-based system was limited by the internet access for data collection. As a solution, we have developed an Android-based mobile application, termed mCHOIS [3]. myCHOIS and its mobile application are now deployed for the DHS in Illinois and for Operation Samahan in California.

myCHOIS has been developed using Open Source Portal Technology with three-tiered Open Grid Services Architecture (OGSA) [1], an accepted standard [45] for accessing Grid and other services under Open Grid Collaborating Environments (OGCE). Users from various School Health Clinics in Illinois are now able to upload data from multiple EMRs into its MySQL database through an Application Programming Interface (API). Reports on aggregated data is generated that can be viewed through web by the authenticated users through a role-based user accessibility. This web portal is designed to insert and upload various data ranging from genomics to socioeconomic, besides the demographic and physiological data. Such data can be analyzed to generate a complete health profile of an individual, which can be used for individualized treatment with the Decision Support System under development. Its web interface is designed to access XSEDE-supported bioinformatics tools that allow a clinician/researcher for 'omics' data analysis. myCHOIS is also designed to provide health educational information and training materials through a *Drupal*-based Content Management System. It also links some useful web sites for research: the National Human Genome Research Institute (NHGRI)'s web site [46] is among those. This site provides a search tool for obtaining genomic information on diseases. Our attempt to search 'Obesity' in this database yielded a set of genes with source of peer-reviewed published information, replication sample size, chromosomal location, strongest SNP-risk allele, risk allele frequency, among others. Such information is useful while creating agent-based model [47] for evidenced-based clinical practice with appropriate decision.

3   C-Grid: The Community Health Grid

As a part of the myCHOIS initiative, we have deployed web-based C-Grid to collect and archive high volume of geographically dispersed data including large video files and other data objects related to healthcare and educational training that can be useful for treating patients at an individualized level. These video files are generated from multiple educational and training events. Similar to ezSRB that was developed earlier [33], we have developed PHP-based ez-iRODS to access and interact with iRODS-server of C-Grid with 5 TB size and RAD 5 configuration. The ez-iRODS is developed as a direct wrapper of the iRODS Client API, PRODS [40]. The C-Grid web portal interface provides input forms for users to interact with iRODS-server through ez-iRODS. The server for C-Grid web portal and the iRODS-server, both located at the San Diego Supercomputing (SDSC), are running with Linux operating system (Fedora). C-Grid server is set up with PHP and MySQL database applications. This server provides functions for storing user accounts, verifying user accounts, communicating with the iRODS-server, and transmitting data objects.

3.1   Design and Development

The C-Grid web portal has been designed to hide the complexity of running an iRODS-server while seamlessly accessing the XSEDE and other resources, such as, resources at the National Center for Biotechnology Information (NCBI). It simplifies the data grid management by employing an effective, flexible, and intuitive user interface that makes presentation simple, enhances the ability of the users to have global access of high volume data in a collaborative environment, and hides the intricacies of the underlying low-level hardware infrastructure from the end-user. The registered users, who are authorized to use this system as collaborators do not need to download, install and plug-in any application programs for accessing the iRODS services. The architecture of C-Grid is shown in Figure 1.

3.1.1 User Interface

The user interface layer of C-Grid web portal acts as multi-platform front-end GUI providing services (e.g., software tools at NCBI, CDC, WHO, etc.) to the users of the Community Health Organizations. Its ez-iRODS component follows Representational State Transfer (REST) [48] architectural style to simplify the service interfaces and to reduce latency. Furthermore, REST architecture allows using HTTP for creating, reading, updating and deleting data operations.

The entire information needed to process a user's request is provided in the form of a Uniform Resource Locator (URL), which identifies a resource (location/data) encompassing headers with query parameters specifying all information about the resource. Communication and collection handling, such as, search and retrieval of the

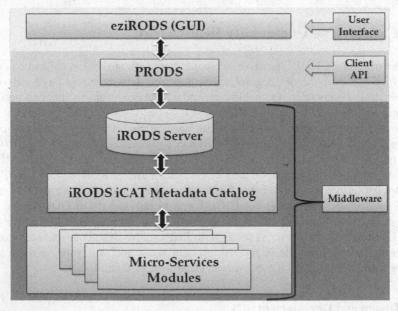

Fig. 1. Architecture of C-Grid web portal. ez-iRODS provides a unified user-friendly interface and conceals the complexity to simplify the process and approach for using the iRODS services.

data is implemented as sequence of *get*, *post*, *delete* and *put* requests. Thus, using stateless REST architecture not only hides the server-side implementation details from the community user, but also from the GUI developer. The users do not have to memorize any commands for interacting with the iRODS-server of C-Grid system.

There are two authentications to support system securities: The first authentication is for accessing the ez-iRODS, and the second authentication is for getting a permission to interact with the iRODS-server. All data objects and collections can be managed and accessed by authorized users through this web-based system. The file "cgrid.php" is the ez-iRODS login web page, which consists of a member authorization method for entering into the system. It also displays a short description of this web-based system and provides two fields for users to enter user's credentials to login into C-Grid web portal. Following code (partially shown) is used for authorizing a user and logging into the system:

```php
<?php
  //add remove files in temp
  $dir = ' ';
  //get directory, first time login, give user default
directory
  if(isset($_GET['dir']) && strlen($_GET['dir'])>0)
    $dir = $_GET['dir'];
  else
    $dir = "/";
  require_once("prods/src/Prods.inc.php");
try {
  // create an iRODS account object for connection by
providing server name, port number, username and password
  $account = new RODSAccount("###.sdsc.edu", ####, "####",
"####");
    //create an dir object, assuming the path is $dir
  $home=new ProdsDir($account,$dir);
    //list home directory
  $childrenDirs=$home->getChildDirs();
  $childrenFiles=$home->getChildFiles();
  //print each child's name
  foreach($childrenDirs as $child) {
  if(strcmp($home->getPath(),'/') !== 0)
          $dirpath = $home->getPath().'/'.$child->getName();
      else
          $dirpath = $home->getPath().$child->getName();
  //print child files
   if(count($childrenFiles) > 0) {
      $color="1";
      foreach($childrenFiles as $child) {
```

```
    $myfile=newProdsFile($account,$home-
>getPath().'/'.$child->getName());
    $myStats= $myfile->getStats();

    <table ...
...  </table>

  <?
  }
} catch (RODSException $exception)
 {echo "Connection failed <br/> \n"
  echo ($exception)
  echo $exception->showStackTrace();
 }
?>
```

After successfully authenticated by its security process, the user is able to interact with the iRODS-server through ez-iRODS for various data grid services. Although every data object is distributed and stored in different locations, the ez-iRODS acts as a virtual file system to *download*, *upload*, *copy*, *create*, *view*, and *remove* data objects, hence facilitating management and access to all collaborative data objects.

Fig. 2. Portal operating menu interface for viewing all collection names in C-Grid. When a user clicks a collection name in the list, the user will be relocated to the subcollection. Then all data objects of the subcollection are displayed on the main menu page for applications. The user also can go back to the current collection or the home collection, which is the starting collection when the user logins to the C-Grid web portal.

The left menu shown in Figure 2 is the user interface for accessing XSEDE resources, communication tools, and other publicly available services. After authentication, the user gets the permission to access to iRODS services shown in the upper menu located in the middle. These include *upload* file, *copy file*, *create collection*, which are important features for utilizing the iRODS services. A user can also register file, register directory, register URL, download file, delete file, preview file, view local collections or network collection list, browse query, move up (in the directory hierarchy), and exit. In addition, some information of each data object, such as the owner name, replica number, location, creation date, object size, and object name can also be viewed through this web page.

3.1.2 Client API
The C-Grid web portal has been designed in PHP and utilizes PRODS as a client access API. PRODS is a PHP client API for iRODS which directly interacts with the iRODS-server. PRODS provides access to iRODS file structure by including a streamer class for native PHP functions that enables PHP core to identify iRODS as legitimate stream. Some of the native PHP functions that support the PRODS streamer class are fopen(), fwrite(), scandir(), and file_get_contents(). PRODS is an open source software package which when downloaded in a local server enables interaction with the iRODS-server via sockets using native iRODS XML protocol.

3.1.3 Middleware
SRB, which was used by us earlier for HASS Grid Project, was robust and scalable in handling scientific data. However, the capability of iRODS to facilitate flexibility and adaptivity in management of data workflows is proved to be superior. In order to enhance our current development and to support future collaborative work, there became a need to add new functionalities, such as, rules mechanism, policy management, etc., to the present data grid system. To achieve these goals as well as simplifying the backend complexity, iRODS became a natural transition when it became available by DICE research group led by Reagan Moore's group [28]. This now serves as a middleware for the C-Grid's development.

When a user performs an action through ez-iRODS, the iRODS-server queries or submits some useful information and the i-commands complete the action from/to an iCAT server (a PostgreSQL server). The iCAT server acts as a catalog to facilitate registering the information about the metadata sets, the users, and the resources managed by the iRODS-server. The resources include data storage instrumentation, supercomputers, local databases and other specialized data storage devices. The iCAT database records the name, type, location and other properties of every resource. In the data grid, the metadata sets are the data (information) related to the data objects of various formats stored in the system. Each data object has associated information in the iCAT database. The information includes the data objects name, a creation date, the owner name, a replica number, a version number, the objects size, the object type, and the resource where the data object is stored.

3.2   Functionality of C-Grid

C-Grid has been developed with the primary objective of collecting and managing the health allied data that comes with multiple different types and formats generated by the participants of the myCHOIS project. This web portal provides powerful mechanism for accruing and exchanging knowledge on health and diseases by offering following core functionalities:

3.2.1   Management and Creation of Collections

C-Grid provides authorized users with a web portal interface to create new collections for storing and managing diverse data objects. The iRODS' collection list interface (shown in Figure 2) not only enables users to create collections but also provides users with the links (URL) to manage their existing collections and corresponding sub-collections. The interface displays various iRODS collections present in a federated network accessible by the authorized users that are located in different zones, e.g., nuZone (for National University) in the SDSC network. An example of the URL (links) provided to the users to access data objects in their collections is shown below:

```
/nuZone/home/nitin/Lsubdir
/nuZone/home/nitin/P subdir/Psubsubdir
```

The above instance shows URL for two collections where nuZone stands for the collection pertaining to the National University zone and /home/nitin corresponds to the home directory of the user (Nitin) under the collection named nuZone. Lsubdir, Psubdir and Psubsubdir represent the subdirectories and sub-collections created by the user (Nitin) under his home directory. A user can move up or down in directory hierarchy structure easily by clicking on the links, which leads user to the respective collections where the user wants to access or manage his/her data objects. Also, a user can create a new collection by providing a parent collection name and a new collection name as inputs to the required fields in the web interface.

3.2.2   Creating, Retrieving and Viewing Files

The C-Grid web portal provides an easy access to accumulate and disseminate the distributed data objects via simple GUI's. Its 'Upload' portal interface (shown in Figure 3) enables community users uploading and copying a local file into the data grid. A user can replicate their files in the same or in another collection using 'copy file' function. This user interface provides functionality for copying or duplicating user's files in the same directory or in other directory belonging to another zone (network), which is accessible to the user. The 'copy file' function requires a user to input the values of Source Object Path Name, Target Object Path Name, and Resource Name in the corresponding fields of 'copy file' web interface. The users can also view the contents of their stored files through the 'view' file GUI. Presently, 'view' function displays the contents if the files are in the text format. If the user has data of different file format in the data grid, the user needs to download the file from the C-Grid through the file 'download' interface to a local computer that has the

Fig. 3. Web-based portal interface for uploading the files in the C-Grid. An upload file web page is for users to upload and to copy a local file into the data grid. The fields such as, iRODS Objects Name, Collection Name, Resource Name, Data Type and Upload File name are necessary for successful uploading action and metadata creation.

needed application program. Furthermore, a 'register directory' web interface link is also provided from the portal's main menu that registers a path of the user's directory in the iCAT server. After the registration, the directory forms a logical sub-collection of a collection which can be assigned by the user as a field input to the iRODS Objects Name in the web interface used for handling data objects.

3.2.3 Handling User-Defined Metadata

The C-Grid web portal allows a user to define metadata for any file or directory stored in the iRODS-server or elsewhere in the network. In addition to the 'upload' file functionality, the GUI (shown in Figure 3) also provides users with the fields to input user-defined metadata that includes information about users, groups, collections, and locations of the data objects. A query in the interface assists users to search a file or a collection based on the user-defined metadata. Metadata search increases the possibility of locating the desired data with the finest recall and accuracy. This search functionality allows users to lookup files using terms or keywords recorded in metadata for the files. Hence, this interface acts as a metadata search engine that receives a metadata query from a user and sends that query to the iCAT server (a PostgreSQL server) for searching a matching file or collection and then returns the file that matches the queried metadata parameters.

3.3   Future Development

Our partner organizations, DHS and Operation Samahan, conduct a significant number of health education events every year. These events are intended for special training to the school children and healthcare providers. We have developed a mobile application based on technology described earlier [3] for an Android-based Smartphone that has been deployed in the field for use by the event coordinators. It is expected that the video files generated from these events will need a large storage space and this C-Grid of 5TB in size will soon be filled. However, this system is scalable. Moreover, video files stored in a multiple video servers can also be networked with C-Grid through ez-iRODS. Nonetheless, federation of this community grid with other health grids, particularly for child health research, will be a powerful resource for the researchers. Its 'admin' module will help in defining rules about data management. Using the rules engine, users can execute rules and microservices to automate the enforcement of management policies including control data access, and manipulation operations locally or at distributed sites as necessary. A mobile application for C-Grid, which is under development, will serve as a useful tool for the 'admin' users to operate the data grid remotely. Further development includes revising this interface to access resources from other grids as shown earlier [50]. C-Grid is designed to serve as a useful tool for the collaborators of myCHOIS Project to operate the data grid remotely.

4   Conclusion

In this paper, we report on a web-based data grid system, termed Community Grid (C-Grid) for community health research, which is developed as a portlet of myCHOIS [1] utilizing grid technology to manage and store health related data. Among the various data management systems, we used iRODS because of multiple advantages as elaborated by Chiang and colleagues [35]. It has been implemented to serve as a distributed computing environment and data management system for sharing resources, data and computing power with the collaborators. It provides the collaborative data sharing and maintenance of distributed storage resource collections. The ez-iRODS of C-Grid conveniently control and interact with the iRODS of the data grid located at the San Diego Supercomputer Center (SDSC). This system utilizing the Data Grid Technologies provides a long-term data preservation and allows user community to access valuable data objects conveniently through the user-friendly intuitive user interface from anywhere. The ez-iRODS simplifies the complicated operating steps and approaches of the iRODS services for the users. This is now deployed for the DHS in Illinois and Operation Samahan in California.

Acknowledgements. The authors acknowledge the suggestion of Arcot Rajasekhar. This project was initiated with technological support of Reagan Moore of SDSC (now at RENCI) and his group, especially to Sheau-yen Chen. Thanks to Peter Dey, and Thomas MacCalla of National University for their support. NS acknowledges the suggestion on codes by Yumiko Iwai of NUCRI. This project is supported in part by the DHS through Coordinated School Health Technical Assistance Grant. C-Grid can be accessed at URL: (http://nucri.nu.edu/cgrid/).

References

1. Datta, A.K., Jackson, V., Nandkumar, R., Sproat, J., Zhu, W., Krahling, H.: CHOIS: Enabling grid technologies for obesity surveillance and control. In: Solomonides, T., Blanquer, I., Breton, V., Glatard, T., Legre, Y. (eds.) Healthgrid Applications and Core Technologies, vol. 159, pp. 191–202. IOS Press, Washington, D.C. (2010)
2. Datta, A.K., Jackson, V., Nandkumar, R., Zhu, W.: Cyberinfrastructure for CHOIS- a global health initiative for obesity surveillance and control. In: Proceedings of the PRAGMA 18, March 3-4 (2010)
3. Datta, A.K., Sumargo, A., Jackson, V., Dey, P.P.: mCHOIS: An application of mobile technology for childhood obesity surveillance. Procedia Computer Science 5, 653–660 (2011)
4. Datta, A.K., Jackson, V., Rimmer, J.: Informatics challenges and opportunities for childhood obesity research. In: Proceedings of Public Health Informatics Conference (PHIN 2011), August 20-24 (2011), http://cdc.confex.com/cdc/phi2011/webprogram/Session12554.html (retrieved on June 4, 2013)
5. Center for Disease Control, http://www.cdc.gov/obesity/data/childhood.html (retrieved on June 4, 2013)
6. Wang, Y., Beydoun, M.A., Liang, L., Caballero, B., Kumanyika, S.K.: Will all americans become overweight or obese? Estimating the progression and cost of the US obesity epidemic. Obesity 16(10), 2323–2330 (2008)
7. Berner, E.S.: Clinical decision support systems: State of the art. ahrq. In: Centers for Disease Control. Publication No.09-0069-EF. Rockville, Maryland: Agency for Healthcare Research and Quality (June 2009)
8. The Health Care Law, http://www.healthcare.gov/law/ (retrieved on June 4, 2013)
9. HIMSS, http://www.himss.org (retrieved on June 4, 2013)
10. Blumenthal, D., Glaser, J.P.: Information technology comes to medicine. New England Journal of Medicine 356(24), 2527–2534 (2007)
11. Hillestad, R., Bigelow, J., Bower, A., Girosi, F., Meili, R., Scoville, R., Taylor, R.: Can electronic medical record systems transform health care? Potential health benefits, savings, and costs. Health Affairs 24(5), 1103–1117 (2005)
12. Homann, L.: Implementing electronic medical records. Communications of the ACM 52(11), 18–20 (2009)
13. The case for Personalized Medicine, http://www.ageofpersonalizedmedicine.org/objects/pdfs/Case_for_PM_3rd_edition.pdf (retrieved on June 4, 2013)
14. Weston, A.D., Hood, L.: Systems biology, proteomics, and the future of health care: toward predictive, preventative, and personalized medicine. Journal of Proteome Research 3(2), 179–196 (2004)
15. Hey, A.J., Trefethen, A.E.: The data deluge: An e-science perspective (2003)
16. Chen, K.Y., Janz, K.F., Zhu, W., Brychta, R.J.: Redefining the roles of sensors in objective physical activity monitoring. Med. Sci. Sports Exerc. 44(1 suppl. 1), S13–S23 (2012)
17. Foster, I., Kesselman, C.: The grid 2: Blueprint for a new computing infrastructure. Morgan Kaufmann (2003)
18. General Parallel File System, http://www-03.ibm.com/systems/software/gpfs/ (retrieved on June 4, 2013)

19. Chervenak, A., Deelman, E., Foster, I., Guy, L., Hoschek, W., Iamnitchi, A., Kessel-man, C., Kunszt, P., Ripeanu, M., Schwartzkopf, B., Stockinger, H., Stockinger, K., Tierney, B.: Giggle: A framework for constructing scalable replica location services. In: ACM/IEEE 2002 Conference Supercomputing, p. 58 (November 2002)
20. Schmuck, F., Haskin, R.: Gpfs: A shared-disk file system for large computing clusters. In: Proceedings of the First USENIX Conference on File and Storage Technologies, pp. 231–244 (2002)
21. Cuff, J.A., Coates, G.M.P., Cutts, T.J.R., Rae, M.: The Ensembl computing architecture. Genome Res. 14, 971–975 (2004)
22. Baru, C., Moore, R., Rajasekar, A., Wan, M.: The sdsc storage resource broker. In: Proceedings of the 1998 Conference of the Centre for Advanced Studies on Collaborative Research, CASCON 1998, p. 5. IBM Press (1998), http://dl.acm.org/citation.cfm?id=783160.783165 (retrieved on June 4, 2013)
23. Rajasekar, A., Moore, R., Wan, M., Schroeder, W., Hasan, A.: Applying rules as policies for large-scale data sharing. In: 2010 International Conference on Intelligent Systems, Modelling and Simulation (ISMS), pp. 322–327. IEEE (2010)
24. Foster, I., Kesselman, C.: Globus: A metacomputing infrastructure toolkit. International Journal of High Performance Computing Applications 11(2), 115–128 (1997)
25. Grimshaw, A.S., Wulf, W.A., et al.: The legion vision of a worldwide virtual computer. Communications of the ACM 40(1), 39–45 (1997)
26. Hoschek, W., Jaen-Martinez, J., Samar, A., Stockinger, H., Stockinger, K.: Data management in an international data grid project. In: Buyya, R., Baker, M. (eds.) GRID 2000. LNCS, vol. 1971, pp. 77–90. Springer, Heidelberg (2000)
27. Huber, V.: UNICORE: A grid computing environment for distributed and parallel computing. In: Malyshkin, V. (ed.) PaCT 2001. LNCS, vol. 2127, pp. 258–265. Springer, Heidelberg (2001)
28. Rajasekar, A., Moore, R., Hou, C.Y., Lee, C.A., Marciano, R., de Torcy, A., Wan, M., Schroeder, W., Chen, S.Y., Gilbert, L., et al.: iRODS Primer: integrated rule-oriented data system. Synthesis Lectures on Information Concepts, Retrieval, and Services 2(1), 1–143 (2010)
29. Southern California Earthquake Center, http://www.scec.org/ (retrieved on June 4, 2013)
30. Biomedical Informatics Research Network, http://www.birncommunity.org/ (retrieved on June 4, 2013)
31. GEON, http://www.geongrid.org/ (retrieved on June 4, 2013)
32. Moore, R.W., Rajasekar, A., Wan, M.: Data grids, digital libraries, and persistent archives: An integrated approach to sharing, publishing, and archiving data. Proceedings of the IEEE 93(3), 578–588 (2005)
33. Franklin, K., Datta, A.K.: Introduction to HASTAC. Cyberinfrastructure for Humanities, Arts Social Sciences Summer Institute. San Diego Supercomputing Center, University of California San Diego. La Jolla, California (July 24-28, 2006)
34. Hedges, M., Blanke, T., Hasan, A.: Rule-based curation and preservation of data: A data grid approach using irods. Future Generation Computer Systems 25(4), 446–452 (2009)
35. Chiang, G.T., Clapham, P., Qi, G., Sale, K., Coates, G.: Implementing a genomic data management system using irods in the wellcome trust sanger institute. BMC Bioinformatics 12(1), 361 (2011), http://www.biomedcentral.com/1471-2105/12/361 (retrieved on June 4, 2013)

36. Hunich, D., Muller-Pfeerkorn, R.: Managing large datasets with iRODS - a performance analysis. In: Proceedings of the 2010 International Multiconference on Computer Science and Information Technology (IMCSIT), pp. 647–654. IEEE (2010).
37. Globus, http://www.globus.org/toolkit/security/index.html (retrieved on June 4, 2013)
38. Kerberos, http://web.mit.edu/kerberos/ (retrieved on June 4, 2013)
39. Jargon, https://www.iRODS.org/index.php/Jargon (retrieved on June 4, 2013)
40. Prods, https://www.irods.org/prods_doc/ (retrieved on June 4, 2013)
41. Pyrods, http://code.google.com/p/irodspython/wiki/PyRods (retrieved on June 4, 2013)
42. iRODS, https://www.irods.org/ (retrieved on June 4, 2013)
43. Health Information Privacy, http://www.hhs.gov/ocr/privacy/ (retrieved on June 4, 2013)
44. Family Educational Rights and Privacy Act (FERPA), http://www.ed.gov/policy/gen/guid/fpco/ferpa/index.html (retrieved on June 4, 2013)
45. Foster, I., Kesselman, C., Nick, J.M., Tuecke, S.: The physiology of the grid. Grid computing: making the global infrastructure a reality, pp. 217-249 (2003)
46. NHGRI, http://www.genome.gov/10001504 (retrieved on June 4, 2013)
47. Hammond, R.A.: Peer reviewed: complex systems modeling for obesity research. Preventing Chronic Disease 6(3) (2009)
48. Fielding, R.T., Taylor, R.N.: Principled design of the modern web architecture. ACM Transactions on Internet Technology (TOIT) 2(2), 115–150 (2002)
49. Salje, E., Artacho, E., Austen, K., Bruin, R., Calleja, M., Chappell, H., Chiang, G.T., Dove, M., Frame, I., Goodwin, A., et al.: Escience for molecular-scale simulations and the eminerals project. Philosophical Transactions of the Royal Society A: Mathematical, Physical and Engineering Sciences 367(1890), 967–985 (1890)
50. Kawai, Y., Iwai, G., Sasaki, T., Watase, Y.: Saga-based application to use resources on dierent grids. In: Proc. International Symposium on Grids and Clouds (ISGC). Proceedings of Science, Taipei, Taiwan (2011)

Local Pre-processing
for Node Classification in Networks
Application in Protein-Protein Interaction

Christopher E. Foley[1,2], Sana Al Azwari[1], Mark Dufton[2],
Isla Ross[1], and John N. Wilson[1]

[1] Department of Computer & Information Sciences
[2] Department of Pure & Applied Chemistry
University of Strathclyde, Glasgow, UK

Abstract. Network modelling provides an increasingly popular conceptualisation in a wide range of domains, including the analysis of protein structure. Typical approaches to analysis model parameter values at nodes within the network. The spherical locality around a node provides a microenvironment that can be used to characterise an area of a network rather than a particular point within it. Microenvironments that centre on the nodes in a protein chain can be used to quantify parameters that are related to protein functionality. They also permit particular patterns of such parameters in node-centred microenvironments to be used to locate sites of particular interest. This paper evaluates an approach to index generation that seeks to rapidly construct microenvironment data. The results show that index generation performs best when the radius of microenvironments matches the granularity of the index. Results are presented to show that such microenvironments improve the utility of protein chain parameters in classifying the structural characteristics of nodes using both support vector machines and neural networks.

1   Introduction

Connected topologies have emerged as a productive way of modelling a wide variety of social, technical and biological systems. Among other domains, the paradigm has been used to characterise social networks [1], protein structures and interactions [3], genetic control [20], market economies [21] and human and machine communication [2]. The power and flexibility of the network concept is highly adaptable as a basis for explaining the overall behaviour of a system but an emerging theme of such modelling is the potential for identifying specific regions in a network that are of particular interest. Such hotspots might represent localised communities in social networks [1] or periods of excessive workload in computer networks [28]. In the remainder of this discussion, we focus particularly on networks that represent protein structural topology and hotspots that characterise points of interaction between proteins. However the novel concept presented here (i.e. the use of localisation to enhance the hotspot detection process) has potential for application in other domains modelled by networks.

M. Bursa, S. Khuri, and M.E. Renda (Eds.): ITBAM 2013, LNCS 8060, pp. 32–46, 2013.
© Springer-Verlag Berlin Heidelberg 2013

The physicochemical properties of proteins provide useful information that results in the identification of new drug targets. Virtual screening offers a methodology for processing entire data collections such as the Protein Data Bank (PDB)[9] with the aim of identifying useful new drug leads. However, it is important to design the screening process in such a way that the maximum benefit is extracted from the data available. An understanding of the structure of proteins and their data representation can guide the design of effective screening methodologies.

A protein is a chain (or combination of chains) of amino acid residues. The chain consists of a backbone that includes the α-carbon atom from each residue in the chain. Protein structures have been modelled as networks with the residues representing the nodes of the graph and edges representing residue interactions [3]. It is also possible to conceptualise nodes as discrete atoms in a protein structure with the edges representing inter-atom factors such as distance. The microenvironment that surrounds each α-carbon is characterised by all the atoms in the network that fall within a defined sphere (Figure 1). Where multiple chains are present, interactions may take place between the residues in different chains. Such protein-protein interaction sites contribute to the function of a protein. Consequently, prediction of the localities of interactions between proteins can be a guide to function prediction for a particular site [11]. Characterising the locality (rather than a point) in a protein structure can be addressed by evaluating microenvironments rather than the specific values associated with particular locations.

Proteins are made from the polymerisation of amino acids into a linear chain that is folded into a three dimensional (3D) structure. The folding pattern brings the functional parts of the protein together and adjusts its configuration in response to binding interactions. The positions of the atoms are determined by processes such as X-Ray Crystallography. Over 70000 3D protein structures are available from the PDB.

A microenvironment is the localised three dimensional spherical neighbourhood surrounding a particular node within a network. In this case the nodes are represented by the α-carbons within a protein structure and the edges by the distance between neighbouring α-carbons. Each microenvironment encloses a variable number of nodes in the network, depending on the radius of the sphere. Figure 2(a) shows the location of amino acids that contribute to the protein-protein interaction site of 1HLU chain A (Bovine Beta-Actin-Profilin). Choosing one particular amino acid (in the case of Figure 2(a) the amino acid at position 89 in the chain is chosen) and defining microenvironments of radius 4Å, 10Å and 20Å causes an increasing number of neighbouring amino acids to be included in the microenvironment cluster as indicated in Figures 2(b) to 2(d). The use of microenvironments as a basis for data mining requires an efficient means of identifying data instances that are within a certain distance of each other. As part of an effort to predict druggable sites on a protein (that is discrete areas where a small drug molecule can regulate the action of the protein) temperature factor can be used as an estimate of flexibility [32]. Protein analysis reveals the variation in 3D location of atoms in a protein as the temperature factor (B factor) parameter. Proteins are not rigid structures. In fact, much of the functionality of a protein depends on small positional adjustments. Temperature factor gives an indication of the likelhood of these adjustments taking place at each atom in the protein structure. The mean temperature factor of all the atoms enclosed in a microenvironment provides a value for this parameter that is based on the flexibility of the locality surrounding a single α-carbon in a chain rather than the flexibility of a single point. This is a consequence of the dependence of the protein topology on the plasticity of the local structure. Since any individual residue may be part of several spheres, pre-processing nodes in the structural network provides an estimate of the protein's behaviour in the surrounding area rather than behaviour at the point represented by each node. This is useful because the activity of a protein is influenced by the general topology rather than point-by-point parameter values, that is in the general context of networks, the structures behave as communities rather than a set of discrete nodes. The temperature factor of a particular residue represents only the flexibility at a specific node in the protein structure network. Other residue parameters such as hydrophobicity[11] (the extent to which the residue repels water) can be evaluated for microenvironments using a similar approach and together these parameters can be used to classify the residues in a chain on the basis of their likely contribution to protein-protein binding sites [10].

A range of processes are available for establishing classifications in datasets, however support vector machines (SVM) and neural networks (NN) typically span the range of prediction accuracies of such methods [10]. An SVM [14]

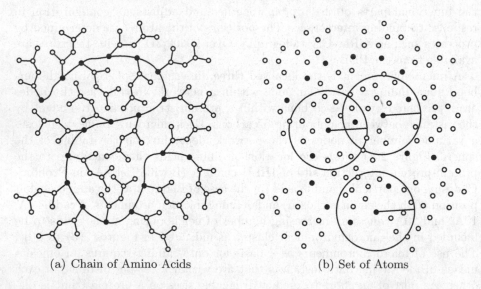

(a) Chain of Amino Acids   (b) Set of Atoms

Fig. 1. Simplified representations of a protein. α-carbons are black and the side chain atoms are hollow. Microenvironment spheres are defined around each of the α-carbons (three are shown).

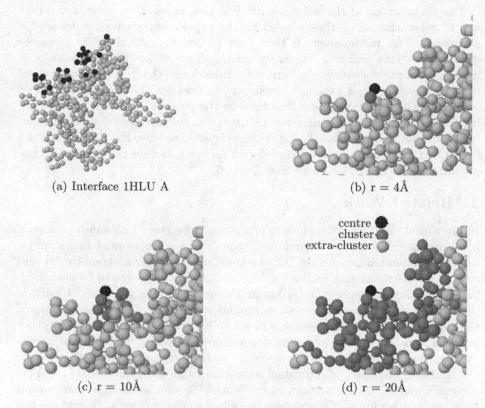

(a) Interface 1HLU A

(b) r = 4Å

centre
cluster
extra-cluster

(c) r = 10Å

(d) r = 20Å

Fig. 2. Amino acid residues clustered in the spheres of varying radii surrounding a single residue centre in the protein-protein interface

is a supervised learning mechanism that generates a hyperplane separating data in a training set. SVMs have been used in bioinformatic research to generate optimal classifications of sites on protein chains [13]. Artificial neural networks represent an alternative approach to classifying input data. They provide a means of deriving a functional model to separate classes in the input data. As with SVMs, NNs are able to classify non-linear data [33]. By training SVMs and NNs with sets of appropriate data, the likely positions of protein-protein interaction sites in a protein chain can be distinguished [24].

There are many possible microenvironment configurations that might be useful in classification nodes within the network. It is not feasible to pre-compute all the possible combinations for any extensive network. Generating microenvironments on-the-fly provides sufficient flexibility and at the same time can support rapid exploration of data. Three dimensional (3D) grid methods have long been known to provide a basis for accelerating the performance of processing spatial data [23]. However in the scenario of varying the level of abstraction of microenvironment data the most appropriate dimensions for box indexes are uncertain.

The contribution of the work described in this paper is to identify the best way of generating an on-the-fly index for the rapid association of nodes within a network. This methodology is then used to demonstrate that the assembly of clustered data makes a significant contribution to predicting hotspots in protein structure networks. In turn, this introduces the general approach of network analysis using localised topological summaries. The rest of the paper is organised as follows: Section 2 establishes the context of related approaches. Section 3 describes index generation and its use in the microenvironment assembly algorithm and presents the experimental work, the results of which are set out in Section 4. The paper concludes with an evaluation of the results and the potential for further work in Sections 5 and 6.

2   Related Work

Improvements in the performance of processing geometric data can be achieved by using specialised data structures. Scenes can be represented in hierarchical trees of bounding volumes [22]. kd-trees are a common structure [8] and their traversal allows intersections to be calculated or distances to be measured. When working with point data, Voronoi diagrams [5] can be used to divide the n-dimensional space into sectors around each point. Using this approach, the entire space within each sector is closer to its parent point than to any other point. This is useful for queries that determine nearest neighbours. In applying these principles to processing molecular data, early recognition of the power of quantising the space of individual molecules came from Leventhal [23]. This approach was further contextualised by Bentley [7] who assumed a quantisation based on search radius. The approach described in the current work explores the assumption that the optimum cell size of the quantisation is the same as search radius. Establishing the optimal approach is an essential step in providing a suitably efficient method of microenvironment assembly.

Residue interaction graphs have been used to characterise protein structures [3] with a view to classifying active sites. Hotspots in both social networks and networks representing protein structures have been found to play a key part in the development and persistence of structural aggregations in their various domains [15]. Network analysis tools have been identified as having considerable potential for identifying targets for drug development [18]. Fixed size microenvironments have been used as a basis for k-means clustering with a view to exploring protein structure [29]. This approach has also been successful in identifying calcium binding sites [6]. The prediction of protein-protein interaction sites has been explored by using combinations of attributes and SVM classifiers [30,11,35]. Typically these lead to prediction accuracy in the region of 60%-70%. Tuning the algorithm by manipulating nearest-neighbour selection produces a prediction accuracy of 73% [31]. Neural networks have also been used to predict protein-protein interaction sites from combinations of physicochemical parameters [17] and are reported to return accuracy in the region of 70%-72%. The use of microenvironments has been found to provide a basis of screening protein data for the presence of allosterically active sites [12].

The novel idea presented here is that pre-processing network data by the construction of parameter aggregates within microenvironments improves the ability to identify hotspot nodes. This contrasts with previous research that focuses on point parameter data. The effectiveness of this approach is demonstrated using both SVM and NN classifiers in the prediction of protein-protein interface sites.

3 Prediction Model

Microenvironment assembly determines the atoms that lie inside the sphere that is centred on each α-carbon in a protein chain. The simple approach of calculating the Euclidean distance between all residue/atom pairs is inefficient in a network that may consist of thousands of nodes. Cell partitioning [7] is used to pre-organise data so that only nearby atoms are considered as candidates during microenvironment assembly. Generation of the cell index is carried out on-the-fly having previously established the overall maxima and minima of spatial coordinates in the collection of protein structures and this is used to produce a 3D grid. The coordinates of each atom associate it with a particular cell in this grid. The index identifies a candidate set of atoms that can be formed from the surrounding cells, immediately ruling out distant atoms from consideration. Only atoms within a reasonable distance of the sphere centre become candidates. The distances between the candidates and the sphere centre are calculated and the appropriate atoms included in the sphere. The index can be tuned by altering the size of the 3D cells and Figure 3 shows how the candidate set is narrowed down by choosing only the cells that intersect with the sphere. Figure 4 gives a formal description of the index generation algorithm. Given the importance of this step in on-the-fly assembly of microenvironments, it is necessary to assess whether the optimal cell edge length is the same as radius size or whether a sub-multiple (L/n) of radius size would be more appropriate.

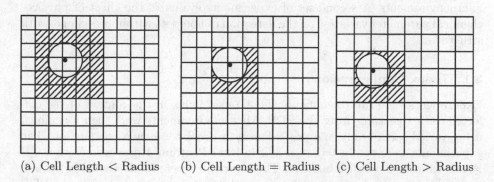

(a) Cell Length < Radius   (b) Cell Length = Radius   (c) Cell Length > Radius

Fig. 3. The relationship between sphere radius and cell length. All of the nearby boxes that intersect with the sphere (or any sphere centred in the same box) are highlighted, and all the atoms from these boxes are candidates for inclusion in the sphere.

```
1:  Create a 3D array of cells encompassing all elements in the PDB.
2:  for each atom in the chain do
3:      Determine which cell this atom belongs in.
4:      Place a reference to the atom in this cell.
5:  for each sphere centre in the chain do
6:      Create an empty sphere.
7:      Determine which cell this sphere centres in.
8:      calculate the length of the shaded area by
```
9: $2 \times \lceil \frac{sphere\ radius}{box\ length} \rceil + 1$
```
10:     for each cell in the shaded area do
11:        for each atom in the cell do
```
12:           distance $= \sqrt{\Delta x^2 + \Delta y^2 + \Delta z^2}$
```
13:           if distance < sphere radius then
14:              add this atom centre to the sphere
```

Fig. 4. 3D grid atom allocation algorithm

At one extreme, a single large cell will place all the atoms together, effectively removing any benefit from the index. At the other end of the spectrum, if the cell size is too small each atom will have its own box, negating the advantage. Figure 3 suggests that choosing a cell size that matches the sphere radius constrains the candidate space but that sub-multiples ($L/2$, $L/3$ etc.) might be more effective. A cell length equal to the microenvironment radius maximally constrains the candidate space volume for the smallest number of candidate cells. This volume can be constrained further by using a greater number of smaller cells. Adding another layer of cells reduces the optimum cell size to $L/2$. A further layer reduces it to $L/3$ and cell lengths for these volume minima can be generalised to L/n.

Experimental work was carried out to evaluate the optimal approach to indexing Protein Data Bank (PDB) data [9] with a view to rapid assembly of microenviroments. A second set of experiments evaluates the effect of physico-chemical parameter variation on the detection of hotspots within protein network structures.

3.1 Index Configuration

Experiments were run to configure the 3D grid index in the context of the collected data structures from the PDB. Microenvironments were then assembled using varied sphere radii and protein sizes in order to allow a comparison. The experiments were conducted on a 3 GHz Intel Pentium 4 processor with 1 GB RAM, running Zenwalk Linux 6.2. The algorithms were implemented in Java 6.

To obtain a representative test dataset, the protein chains present in the PDB were divided into groups by chain length (1–50 residues in the first group, 51–100 in the second, etc.) and one chain was chosen at random from each group. The final dataset is shown in Table 1.

Table 1. Set of protein chains used for benchmarking the index algorithms

ID	Ch.	Len	ID	Ch.	Len	ID	Ch.	Len	ID	Ch.	Len
1D9M	A	18	1AZP	A	66	2I8T	B	149	2IX3	B	972
1BGL	F	1021	3DEE	A	197	1J2Q	B	223	2QPQ	C	296
2QN1	A	813	2VC9	A	882	1MIQ	B	327	2OF6	B	400
1JRP	G	450	1UYT	A	681	1N7O	A	721	2HLD	S	480
1ZPU	E	529	1EFK	A	553	2PPB	M	1119	1WZ2	B	948
2AHX	B	615	1JRP	B	760						

The performance evaluation of microenvironment assembly was carried out by repeating the algorithm 1000 times. In order to make sure the compiled and optimised execution was measured, the 1000 measurements were repeated until two consecutive measurements were within 10% of each other. To prevent the benchmarked code being optimised out by the compiler, a summation of the sphere results was calculated and output to the console after the time measurements were complete. To determine the best cell size, the protein size was kept constant. Chain E from 1ZPU was chosen since it is in the middle of the range of chain length, which fixed the number of residues at 529.

An index using cells that are too small will take a long time to create while very large cells will approach $O(n^2)$ in terms of microenvironment assembly performance. Somewhere between these two extremes must lie the maximum efficiency. The experiment was run at sphere radii of 4, 5, 6, 7, 8, 9 and 10Å. For each sphere size, the cell size was varied from 4 Å to 20 Å in steps of 1Å. The best cell size was deduced from the above experiments and used to benchmark the cell index at sphere radii of 4, 5, 6, 7, 8, 9 and 10Å for each chain length in the dataset.

3.2 Aggregating Parameter Values

Having established the optimal approach to microenvironment assembly, experiments were conducted to examine the effect of the approach on the classification of nodes within the protein structural network. In this case, the intention was to identify hotspot nodes representing those residues that take part in protein-protein interactions. Figure 5 illustrates the basis of this approach. In this case, temperature factor is shown for each position in a sample chain (1HLU chain A). The sites of the protein-protein interface are indicated and can be seen to be distributed over peaks and troughs when using a 0Å sphere (i.e point data). In the case of the 40Å dataset, the protein-protein interaction sites have coalesced in troughs in the distribution. This suggests that particular features of the temperature factor distribution are evident at some microenvironment radii but not at others. These variations in parameter distributions are of use in classifying nodes in the network structure of residues.

Table 2. Sample dataset used for benchmarking the microenvironments

ID	Chs.	ID	Chs.	ID	Chs.	ID	Chs.	ID	Chs.	ID	Chs.	ID	Chs.	ID	Chs.
1AHW	B C	1AVG	H I	1AY7	A B	1AZS	C B	1B6C	A B	1B7Y	A B	1BDJ	A B	1BI7	A B
1BMQ	A B	1BP3	A B	1BVK	A C	1BVN	P T	1D4V	A B	1DAN	L U	1EBD	A C	1EFU	A B
1ETH	A B	1GFW	A B	1GLA	F G	1GOT	B G	1HJA	B I	1HLU	A P	1IRA	X Y	1KKL	A H
1L0Y	A B	1NOC	A B	1PDK	A B	1QBK	B C	1SMP	A I	1STF	E I	1UDI	E I	1UEA	A B
1VAD	A B	1ZBD	A B	2PCC	A B	3EZE	A B	7CEI	A B						

Table 3. Dataset parameters

Parameter	Characteristic
Temperature factor (B-factor)	The flexibility of the protein at a particular atom.
Druggability	The likelihood of targetting by a drug-like molecule.
Hydrophobicity	The extent to which the residue repels water.
Total atomic weight	The local size at a node.
Residue number	Position in the protein chain.

Table 4. Dataset size

Parameter	Count
Total residues	24526
Exposed residues	7977
Interface residues	4104

To verify this assumption, a set of chain pairs with prior established protein-protein interaction sites was identified [4]. This set was further refined by removing chains containing multiple models (i.e. where variations in the configuration of the protein were possible) and those identified as containing significant redundancy [24]. Lastly, chains that had no matching pair subsequent to these steps were also removed. Following this process, the sample set consisted of those chain pairs shown in Table 2.

The residues in each chain of this set were then classified on the basis of their proximity to residues in the complementary chain. The occurrence of two α-carbons from complementary chains within a range of 12Å was taken as an indication that the respective locations of these residues represented a protein-protein interaction site [24]. The α-carbons in each chain were also classified on the basis of their accessible surface area (ASA) [19]. Surface residues were taken to be those with an ASA more than 20% of the surface area. The dataset chosen is summarised in Table 4.

A set of parameters indicated by previous work [16,26] was then derived for each surface residue in each chain. The parameters represent orthogonal characteristics of nodes within the network as shown in Table 3. Microenvironments of radii 0Å to 50Å were used to produce a mean value for each of these parameters for each node using the approach explained in subsection 3.1. For temperature factor and total atomic weight, each atom in the microenvironment provides a contribution to this mean. For other parameters used, each residue included in the locality provides a contribution.

This approach generated a vector of values that were used with both neural net (NN) and support vector machine (SVM) classifiers. Non-overlapping training and test sets were generated by assigning alternate microenvironments to each of these two sets. The LIBSVM package [13] was used to develop a support vector machine classification model using the training set of residues. In the course of generating the classification, cross-validation was carried out within the training set. A non-linear function provided the best separation for distinguishing protein-protein interaction sites using an SVM. Neural net classifiers were generated and tested using Matlab [25] and the same training and test sets as were used to generate the SVM classifications. As with the SVM classifiers, cross-validation was carried out within the training set during generation of the NN classifiers.

4   Experimental Results

The performance of microenvironment assembly based on a properly configured cell index is shown in Figure 6. The maximum number of amino residues in a single chain in the PDB is 4128, suggesting that index generation will be around 0.6 seconds in the worst case.

It can be seen from Figure 7, that the most efficient cell size is equal to the sphere radius. As the cell size increases from this global minimum, the time for the algorithm to run increases steadily. This is consistent with the larger cells holding progressively more atoms and therefore requiring more distance calculations. As the cell size decreases from the global minimum, the trend is for the time to increase. This is because more cells are required and their creation becomes the most time-intensive step. However Figure 7 also shows local minima at half the optimum cell size. Consider determining the sphere at a 7Å radius. When the cell size is also 7Å, the candidate list is drawn from the central cell and all of the surrounding cells. If the cell size is decreased to 6Å we still have to check the central cell and the surrounding ones. However, now the range of the sphere can include atoms up to two cells away. If we go below 3.5Å (half the optimum box size), we have to consider atoms three cells away. One could expect another local minimum at 1.75Å, another at half this, and so on. The results from the optimum configuration are shown in Figure 8. Microenvironment assembly using the 3D grid index differs in that the speed varies with the sphere size.

The impact of changing the radius of the sphere on the identification of hotspot nodes in the context of their contribution to protein-protein interfaces is shown in Figures 9 and 10. The precision and recall of both SVM and NN show a gradual increase over the sphere radius from 0Å to 40Å. This variation can be seen more clearly in the context of the prediction accuracy shown in Figure 10. The SVM approach shows a peak accuracy of about 80% occuring at a radius of 40Å. NN accuracy also peaks at the same radius. To explore the distribution of data contributing to these predictions, Figure 11 shows the coefficient of variation (the ratio of standard deviation to mean) for each parameter over the radii chosen. Figure 12 shows the impact of isolating the contribution of each parameter to the SVM prediction. Here the microenvironment radius was fixed at 40Å except for the indicated parameter, which was varied in the range 0Å to 50Å.

At lower microenvironment radii, the temperature factor provides the dominant component of the overall accuracy of the predictive model. This covers increasing radii upto about 30Å. Between 30Å and 50Å other parameters,

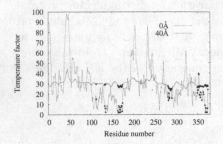

Fig. 5. Residues contributing to the protein-protein interface superimposed on the temperature factor distribution for 1HLU A

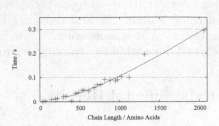

Fig. 6. Summarisation using the 3D grid index

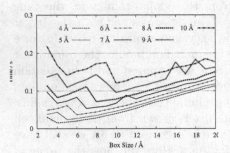

Fig. 7. Effect of cell size on execution time for different sphere radii

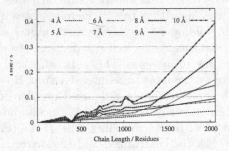

Fig. 8. Index generation at different sphere radii. Cell sizes are set equal to the sphere radius.

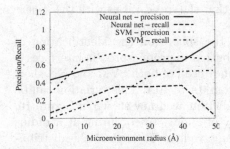

Fig. 9. Precision and recall at varying microenvironment radii

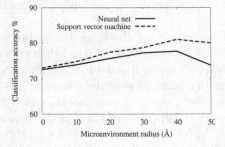

Fig. 10. Accuracy at varying microenvironment radii

particularly hydropathy play an increasing role in contributing towards the accuracy of the prediction. In all cases however, the use of microenvironments as the basis of generating classifications show evident improvements over the use of data points that do not take into account the neighbourhood surrounding the α-carbon.

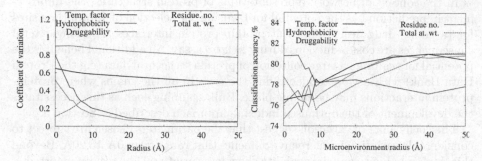

Fig. 11. Coefficients of variation   **Fig. 12.** Principal component analysis

5 Discussion

The experimental work verifies the assumption in constructing spheres in network structures, that the most appropriate cell length matches sphere radius. This result provides confidence in the optimal performance of microenvironment assembly, which is necessary for locating good classifiers within the search space. The on-the-fly approach is efficient enough to remove the necessity for materialising microenvironments. This improves the utility of the method since it removes the need for predicting the combinations of parameters that can distinguish hotspot nodes. If a user interface must respond to a mouse click or a keystroke within 0.1 s, the 3D grid index continues to meet the criteria up to about 1200 residues for the higher sphere sizes and over 2000 residues for sphere sizes of 7Å and under. This makes it feasible to build a direct manipulation interface to large networks such as those represented by the PDB and provides support for interactive data mining. The experiments show that the performance of the index depends on sphere size, with larger radii making the index less efficient.

Sphere size also has an effect on the use of microenvironments as a basis for classifying hotspots in the network, in this case characterising residues in terms of their contribution to protein-protein interface sites. Microenvironments provide a more effective quantification of the impact of parameter values at a particular site than is available by using localised point values. This effect is visible in predictions based on SVMs and NNs using the same set of parameter values. The benefit of using microenvironments as input into SVMs is significant. The accuracy of the prediction reaches 80% at a sphere radius of 40Å. This compares favourably with results in the range 60%-73% that are reported in the literature [30,11,35,31]. Using network centrality analysis on residue interaction graphs predicts active sites with a accuracy of 70% [3]. An equivalent effect is noticeable in the precision and recall of SVMs operating on the test set. Precision and recall at 40Å are 63% and 65% respectively, compared with prior reported equivalent values of 46% and 67% [30]. In this context, neural networks return lower values for precision, recall and accuracy, a point already noted in classifiers developed to address other domains [10].

In the context of network representations of protein structures, the improvement in prediction has the potential for focusing the selection of appropriate sites for targeting drug design aimed at protein-protein interfaces. The longer term consequences are cost reductions during *in vitro* assay. An additional benefit is the potential for scanning a large collection of protein structural data (e.g the Protein Data Bank) with a view to identifying the sites in all the chains where protein-protein interactions may be taking place. Bulk scanning such as this necessitates the development of the optimised indexing approach described in Section 3.

The coefficients of variation suggest that the parameters chosen are subject to considerable variation in microenvironments that range from 0Å to 20Å. Beyond this, hydrophobicity and druggability show less variation. The results reported in Figure 12 suggest that despite restricted variation of these parameters beyond 20Å, they still make an important contribution to the prediction process at microenvironment radii of around 40Å.

This work has focused on the use of 3D coordinates to model protein structures as an example of nodes located within a network. Other approaches to protein modeling include representation as graph structures with edges typically denoting the spatial proximity of atoms within the structure [27]. In this idiom, microenvironments are an appropriate way of characterising the physicochemical topology of proteins because they can be parameterised to span variable sub-graphs within the chain. The utility of this approach is demonstrated in the increased classification accuracy for microenvironment centres. Other applications of graph theory include analysis of social network activity, ecological systems and economic structures [1]. Within such domains, there is considerable challenge in the identification of communities as collections of interconnected nodes [34]. Search methodologies can be deployed to address this problem but they are typically limited in the range of network sizes that can be analysed. Microenvironments are an appropriate tool that can be applied in this context and we are currently developing our approach in this direction.

6   Conclusion

The experimental work reported has evaluated the efficiency of a parameterised 3D grid index for generating microenvironment data for use in the classification of residues in terms of their contribution to protein-protein interface sites. The index was evaluated with protein atomic coordinates and has been shown to be most efficient when the cell size matches the granularity of the summary.

The optimised approach to indexing provides a basis for bulk scanning of protein data to identify sites where protein-protein interactions may occur. The use of microenvironments rather than underlying point data values provides a basis for improving the classification performance of both SVMs and NNs in exploring protein structures. Prediction accuracy increases progressively up to about 80% at a microenvironment radius of around 40Å. The model of node classification based on microenvironments in protein network structures has potential for application in other domains where network size makes conventional analysis infeasible.

References

1. Aggarwal, C.: Social Network Data Analytics. Springer (2011)
2. Ahlswede, R., Cai, N.C.N., Li, S.Y.R., Yeung, R.W.: Network information flow (2000)
3. Amitai, G., Shemesh, A., Sitbon, E., Shklar, M., Netanely, D., Venger, I., Pietrokovski, S.: Network analysis of protein structures identifies functional residues. J. Mol. Biol. 344(4), 1135–1146 (2004)
4. Ansari, S., Helms, V.: Statistical analysis of predominantly transient protein-protein interfaces. Proteins 61, 344–355 (2005)
5. Aurenhammer, F.: Voronoi diagrams-a survey of a fundamental geometric data structure. ACM Comput. Surv. 23, 345–405 (1991)
6. Bagley, S., Altman, R.: Characterizing the microenvironment surrounding protein sites. Protein Science 4, 622–635 (1995)
7. Bentley, J., Stanat, D., Hollins Williams, E.: The complexity of finding fixed-radius near neighbors. Information Processing Letters 6(6), 209–212 (1977)
8. Bentley, J.L.: Multidimensional binary search trees used for associative searching. Commun. ACM 18, 509–517 (1975)
9. Berman, H.M., et al.: The Protein Data Bank. Acta Crystallogr. D 58(6, pt. 1), 899–907 (2002)
10. Bisbal, J., Engelbrecht, G., Villa-Uriol, M.-C., Frangi, A.F.: Prediction of cerebral aneurysm rupture using hemodynamic, morphologic and clinical features: A data mining approach. In: Hameurlain, A., Liddle, S.W., Schewe, K.-D., Zhou, X. (eds.) DEXA 2011, Part II. LNCS, vol. 6861, pp. 59–73. Springer, Heidelberg (2011)
11. Burgoyne, N.J., Jackson, R.M.: Predicting protein interaction sites: binding hotspots in protein-protein and protein-ligand interfaces. Bioinformatics 22(11), 1335–1342 (2006)
12. Foley, C.E., AlAzwari, S., Dufton, M., Wilson, J.N.: Using microenvironments to identify allosteric binding sites. In: Proc. IEEE International Conference on Bioinformatics and Biomedicine, pp. 1–5 (2012)
13. Chang, C., Lin, C.: LIBSVM: A library for support vector machines. ACM TOIST 2, 27:1–27:27 (2011)
14. Cortes, C., Vapnik, V.: Support-vector networks. Machine Learning 20, 273–297 (1995)
15. Csermely, P.: Creative elements: network-based predictions of active centres in proteins and cellular and social networks. Trends in Biochemical Sciences 33(12), 569–576 (2008)
16. Ezkurdia, I., Bartoli, L., Fariselli, P., Casadio, R., Valencia, A., Tress, M.L.: Progress and challenges in predicting protein-protein interaction sites. Briefings in Bioinformatics 10(3), 233–246 (2009)
17. Fariselli, P., Pazos, F., Valencia, A., Casadio, R.: Prediction of protein-protein interaction sites in heterocomplexes with neural networks. European Journal of Biochemistry 269(5), 1356–1361 (2002)
18. Farkas, I.J., Korcsmaros, T., Kovacs, I.A., Mihalik, A., Palotai, R., Simko, G.I., Szalay, K.Z., Szalay-Beko, M., Vellai, T., Wang, S., Csermely, P.: Network-Based Tools for the Identification of Novel Drug Targets. Sci. Signal. 4(173), pt. 3+ (2011)
19. Kabsch, W., Sander, C.: Dictionary of protein secondary structure: pattern recognition of hydrogen-bonded and geometrical features. Biopolymers 22, 2577–2637 (1983)

20. Kauffman, S.A.: Metabolic stability and epigenesis in randomly constructed genetic nets. Journal of Theoretical Biology 22(3), 437–467 (1969)
21. Kirman, A.P.: The economy as an evolving network. J. Evolutionary Economics 7(4), 339–353 (1997)
22. Klosowski, J., Held, M., Mitchell, J., Sowizral, H., Zikan, K.: Efficient collision detection using bounding volume hierarchies of k-dops. IEEE T. Vis. Comput. Gr. 4(1), 21–36 (1998)
23. Levinthal, C.: Molecular model-building by computer. Scientific American 214, 42–52 (1966)
24. Liu, R., Jiang, W., Zhou, Y.: Identifying protein-protein interaction sites in transient complexes with temperature factor, sequence profile & accessible surface area. Amino Acids 38(1), 263–270 (2010)
25. MATLAB. version 7.13.0 (R2011b). The MathWorks Inc., Natick, Massachusetts (2011)
26. Shinji, S., Hiroki, S., Kobori, M., Noriaki, H.: Use of amino acid composition to predict ligand-binding sites. J. Chem. Inf. Model. 47, 400–406 (2007)
27. Vishveshwara, S., Brinda, K., Kannan, N.: Protein structure: insights from graph theory. Journal of Theoretical and Computational Chemistry 1(1), 1–25 (2002)
28. Wood, T., Shenoy, P., Venkataramani, A., Yousif, M.: Sandpiper: Black-box and gray-box resource management for virtual machines. Computer Networks 53(17), 2923–2938 (2009)
29. Wu, S., Liu, T., Altman, R.: Identification of recurring protein structure microenvironments and discovery of novel functional sites around cys residues. BMC Struct. Biol. 10(4) (2010)
30. Xia, J., Zhao, X., Song, J., Huang, D.: Apis: accurate prediction of hot spots in protein interfaces by combining protrusion index with solvent accessibility. Bioinformatics 11(174), 1–14 (2010)
31. Gui, J., Yang, L., Xia, J.F.: Prediction of protein-protein interactions from protein sequence using local descriptors. Protein Pept. Lett. 17(9), 1085–1090 (2010)
32. Yuan, Z., Bailey, T.L., Teasdale, R.D.: Prediction of protein B-factor profiles. Proteins: Struct., Funct., Bioinf. 58(4), 905–912 (2005)
33. Zhang, G.: Neural networks for classification: A survey. IEEE Transactions on Systems, Man and Cybernetics - Part C 30(4), 451–462 (2000)
34. Zhao, Y., Levina, E., Zhu, J.: Community extraction for social networks. Proc. National Academy of Sciences 108(18), 7321–7326 (2011)
35. Zhong-Hua, S., Fan, J.: Prediction of protein binding sites using physical and chemical descriptors and the support vector machine regression method. Chinese Physics B 19(11), 110502 (2010)

Automatic Evaluation of FHR Recordings
from CTU-UHB CTG Database

Jiří Spilka[1], George Georgoulas[2], Petros Karvelis[2], Vangelis P. Oikonomou[2],
Václav Chudáček[1], Chrysostomos Stylios[2], Lenka Lhotská[1], and Petr Janků[3]

[1] Dept. of Cybernetics, Czech Technical University in Prague, Czech Republic
[2] Dept. of Informatics and Communications Technology, TEI of Epirus, Arta, Greece
[3] Dept. of Gynecology and Obstetrics,
Teaching Hospital of Masaryk University in Brno, Czech Republic
spilka.jiri@fel.cvut.cz

Abstract. Fetal heart rate (FHR) provides information about fetal
well-being during labor. The FHR is usually the sole direct information channel from the fetus – undergoing the stress of labor – to the
clinician who tries to detect possible ongoing hypoxia. For this paper, new
CTU-UHB CTG database was used to compute more than 50
features. Features came from different domains ranging from classical
morphological features based on FIGO guidelines to frequency-domain
and non-linear features. Features were selected using the RELIEF
(RELevance In Estimating Features) technique, and classified after
applying Synthetic Minority Oversampling Technique (SMOTE) to the
pathological class of the data. Nearest mean classifier with adaboost
was used to obtain the final results. In results section besides the direct
outcome of classification the top ten ranked features are presented.

Keywords: fetal heart rate, intrapartum, feature selection,
classification.

1 Introduction

Electronic fetal monitoring (EFM) is used for fetal surveillance during
pregnancy and, more importantly, during delivery. The EFM most commonly
refers to cardiotocography (CTG) that is a measurement of fetal heart rate
(FHR) and uterine contractions (UC). Since its introduction the CTG has served
as the main information channel providing obstetricians with insight into fetal
well-being. CTG monitoring still plays a role of the most prevalent method
in use for monitoring of antepartum as well as intrapartum fetal well-being.
The goal of fetal monitoring is to prevent fetus of potential adverse outcomes
and provide an information about his/hers well-being. The main advantage of
CTG, when compared to previously used auscultation technique, lies in its
ability of continuous fetal surveillance though, this advantage is claimed to
be insignificant in preventing adverse outcomes (with exception of neonatal
seizures) as described in meta-analysis of several clinical trials [1]. The other

M. Bursa, S. Khuri, and M.E. Renda (Eds.): ITBAM 2013, LNCS 8060, pp. 47–61, 2013.
© Springer-Verlag Berlin Heidelberg 2013

main controversies of CTG include: increased rate of cesarean sections [1] and high intra- and inter-observer variability [2,3].

Nowadays CTG remains the most prevalent method for intrapartum fetal surveillance [2,4], often supported by ST-analysis (Neoventa Medical, Sweden) which is based on analysis of fetal electrocardiogram (FECG). The introduction of additional ST-analysis into the clinical practice improved the labor outcomes slightly [5,6] but its use is not always possible or feasible since it requires invasive measurement. Moreover, in order to use ST-analysis the correct interpretation of CTG is still required.

The interpretation of CTG is based on FIGO guidelines [7] introduced in 1986, or their newer international alternatives [8]. The main goal of guidelines is to assure lowering of the number of asphyxiated neonates while keeping the number of unnecessary cesarean sections (due to false alarms) at possible minimum. Additional goal of the guidelines was to lower the high inter and intra-observer variability. Despite the efforts made, the variability of clinicians evaluation of CTG still persists [9]. Three possible ways to lower it were discussed. e.g. [10] i) by extensive training, ii) using the most experienced clinician as an oracle, iii) and/or by computerized system supporting clinicians with the decision process.

The attempts of computerized CTG interpretation are almost as old as the FIGO guideline themselves. Beginning with work of [11] the automatic analysis of CTG was aligned with clinical guidelines [12]. Beyond the morphological features used in the guidelines, new features were introduced for FHR analysis. These were mostly based on the research in the adult heart rate variability [13]. The statistical description (time domain) of CTG tracings was employed in [14] and in [15]. The spectrum of FHR either in antepartum or intrapartum period offered insight to fetal physiology, and the short review [16] described recent development in this area. The joint time-frequency analysis of FHR in the form of wavelet analysis was employed in [17]. Nonlinear methods are widely used for FHR analysis [18,19] and in our recent work we showed their usefulness in this field [20]. Different approaches were used for classification of FHR into different categories either based on pH levels, base deficit, or other clinical parameters. These approaches includes: Support Vector Machines (SVMs) [17,21,20], artificial neural networks (ANNs) [22,23], or a hybrid approach utilizing grammatical evolution [24].

The contributions of the paper are twofold: First, from the CTG point of view, the used database will be open access at the time of publication, This is one of the largest databases used for automatic evaluation of the CTG. Second, we provide a promising approach for the automatic classification of CTG using the umbilical pH value as a gold standard. The results could serve as a base methodology for a new algorithm development on clinically sound data. An overview of the procedure is shown in Fig. 1.

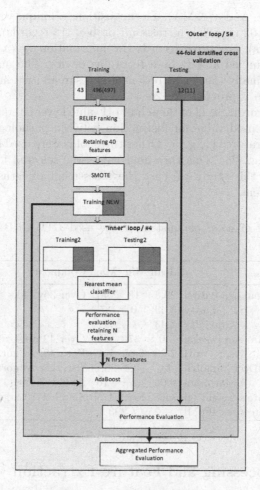

Fig. 1. An overview of the procedure

2 Data Used

The database of 552 records is a subset of 9164 intrapartum CTG recordings that were acquired between years 2009 and 2012 at the obstetrics ward of the University Hospital in Brno, Czech Republic. The CTG signals were carefully selected with clinical as well as technical considerations in mind. The main parameters and their distributions are presented in Table 1. We have decided to select recordings that ended as close as possible to the birth and that had in the last 90 minutes of labor at least 40 minutes of usable signal. Additionally since CTG signal at II. stage of labor is very difficult to assess [25], we have included to the database only those recordings which had II. stage at maximum 30 minutes-long. The CTGs were recorded using STAN S31 (Neoventa Medical, Mölndal, Sweden) and Avalon FM40 and FM50

(Philips Healthcare, Andover, MA). The acqusition technique was either by scalp electrode (FECG 102 records), ultrasound probe (412 records), or combination of both (35 records). For three records the information was not available. All recordings were sampled at 4Hz by a recording device. The majority of babies were delivered vaginally (506) and rest using caesarean section (46). The more detailed description is provided in [26].

From the 552 recordings, 44 of them had pH value lower or equal to 7.05 as this is most commonly used value for distinction between pathological and normal outcome in the literature cf. e.g. [5]. Other thresholds were used in research work, for overview see e.g. [26]. This threshold value was selected for the formation of two classes. In this study we cast the assessment of fetus wellbeing as a classification problem.

Table 1. Overview of main parameters of the used CTU-UHB cardiotocography database

	Mean	Min.	Max.	Comment
Maternal age [years]	29,8	18	46	Over 36y: 40.
Parity	0,43	0	7	
Gravidity	1,43	1	11	
Gesta. age [weeks]	40	37	43	Over 42 weeks: 2
pH	7,23	6,85	7,47	Pat.: 48; Abnor.: 64
BDecf [mmol/l]	4,6	-3,4	26,11	Pat.: 25; Abnor.: 68
Apgar 5min	9,06	4	10	AS5 < 7: 50
Neonate's weight [g]	3408	1970	4750	Small: 17; Large: 44
Neonate's sex [F/M]	259 / 293			

3   Signal Processing and Feature Extraction

3.1   Signal Preprocessing

The FHR was measured either externally using Doppler ultrasound (US) or internally by a scalp electrode (DECG); in special cases the combination of methods was used, i.e. beginning recording with US measurement and ending with DECG measurement. FHR recorded externally has lower signal to noise ratio than that recorded internally. The artifacts could be caused by mother/fetal movement, displacement of ultrasound probe, or simply by mis-detection of fetal heart beat by the recording device. We employed a simple artifacts rejection scheme: let $x(i)$ be a FHR signal in beats per minute (bpm), where N is number of samples and $i = 1, 2, \ldots, N$, whenever $x(i) \leq 50$ or $x(i) \geq 210$ we interpolated $x(i)$ using cubic Hermite spline interpolation. We used interpolation implemented in MAT-LAB®. We interpolated artifacts or missing data when the length of missing signal was equal or less than 15 seconds – the value based on FIGO guidelines and our experiments. When computing features we skipped the long gaps (> 15 seconds). An example of the result of artifacts removal is presented in Fig. 2.

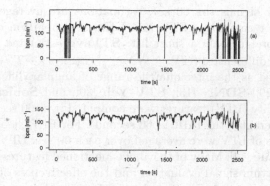

Fig. 2. Artefacts rejection. (a) Raw signal with artefacts, (b) signal after artefacts rejection.

3.2 Feature Extraction

As mentioned above the FIGO guidelines features were essential for the development of any system for automatic classification. Beyond that, other features, originating from different domains, were examined and used for classification. In this section we briefly describe the features, the description should serve as a context necessary to reproduce the analysis. We refer the interested reader to the referenced papers or to our previous works [20,27].

Morphological features (used in clinical settings). Morphological features proposed in the FIGO guidelines represents macroscopic – "visible" – properties of the FHR. The morphological features were as follows: **mean of FHR baseline**, where the baseline is the mean level of fetal heart rate where acceleration and deceleration are absent; **number of accelerations**, where acceleration is a transient increase in heart rate above the baseline by 15 bpm or more, lasting 15 seconds or more; **number of decelerations**, where deceleration is a transient episode of slowing fetal heart rate below the baseline level by more than 15 bpm and lasting 10 seconds or more.

Short/long term variability. The short term variability (STV) is the only feature sometimes computed automatically in clinical settings. The computation of STV depends whether FHR is recorder internally or externally. For the internal recording real beat-to-beat variability could be estimated while for external monitoring there is no real beat-to-beat (BB) variability because of intrinsic smoothing due to the correlation based technique. Instead epoch-to-epoch (EE) variation is used when the FHR is averaged over short period of time (2.5-3.75 sec.). Recall that $x(i)$ is the i-th FHR sample in beat per minute (bpm), let $T(i)$ a FHR sample in milliseconds $i = 1, 2, \ldots, N$, where N is the length of FHR. As noted in [28] the STV computed using $x(i)$ and $T(i)$ is not always the same because of dependence on the value of FHR mean utilized in some variability computation. The STV is estimated for signals of length

60 sec.; for longer signals the 60 sec. estimations are averaged. There exist several methods for computing STV and LTV, a comparison could be found in [28]. Here we present only a short list: **STVavg** estimated as the average of successive beat differences: $\text{STV} = \frac{1}{N}\sum_{i=1}^{N-1}|T(i+1) - T(i)|$ [ms], **STV-DeHann** estimated as the inter quartile range of angular differences between successive $T(i)$s [29], **SDNN** [13], **STV-Yeh** [30], and **Sonicaid 8000** [31]. Long term variability (LTV) features were computed over 60 seconds and there was no need of averaging the FHR in 60 seconds. For FHR signals longer than 60 sec. estimations of LTV were averaged over each 60 sec. **LTV-DeHaan** [29] and the **Delta value** [14]. Many of the above mentioned features have been used in cases of antepartum signal evaluation and the effectiveness of many of them depends on their performance in the presence of accelerations and decelerations.

Frequency domain features. Various spectral methods have been used for the analysis of adult heart rate [13]. In the case of FHR analysis, no standardized use of frequency bands exists. Therefore we used two slightly different partitionings of the frequency bands as was previously used in our work [21]. First we divided the frequency range into 3 bands [13] and calculated the energy of the signal in each one of them: **Very Low Frequency** (VLF); **Low Frequency** (LF) referred to as Mayer waves and **High Frequency** (HF) corresponding to fetal movement. Additionally the **ratio of energies** in the bands: $\text{ratio\_LF\_HF} = \frac{LF}{HF}$ was computed. It is a standard measure in adults and expresses the balance of behavior of the two autonomic nervous system branches. The alternative frequency partitioning followed suggestions of [32]. They proposed the following 4 bands: **Very Low Frequency** (VLF); **Low Frequency** (LF) correlated with neural sympathetic activity; **Movement Frequency** (MF), related to fetal movements and maternal breathing; **High Frequency** (HF), marking the presence of fetal breathing. Similarly to the previous 3-band division the following ratio of energies was computed: $\text{ratio\_LF\_MFHF} = \frac{LF}{MF+HF}$. This ratio is supposed to quantify the autonomic balance control mechanism (in accordance with the LF/HF ratio normally calculated in adults). The spectrum of FHR was estimated using the fast Fourier transform.

Nonlinear features. Almost all nonlinear methods used for FHR analysis have their roots in adult HRV research. For nonlinear features we detrened FHR by the estimated baseline and also normalized the signal to have zero mean and unit variance. The **Poincaré plot** is a basic nonlinear feature commonly used in HRV domain [13]. The plot is a geometric representation of HRV where each RR interval is plotted as a function of the previous one. In this work we estimated waveform fractal dimension by several methods. These were: **box-counting dimension**, which expresses the relationship between the number of boxes that contain part of a signal and the size of the boxes; **the Higuchi method (FD_Hig)** [33], where the curve length $\langle L(k) \rangle$ is computed for different steps k and it is related to the fractal dimension by an exponential formula; the **variance fractal dimension (FD_Var)** that is based on properties of fractional Brownian motion. The variance $sigma^2$ is related to the time increments

Δt of a signal $X(t)$ according to the power law [34]; an estimate of the fractal dimension proposed by **Sevcik** [35]; Detrend Fluctuations Analysis (DFA) [36] for estimating the fractal dimension, D, via scaling exponent α, $D = 3 - \alpha$. For all methods, the fractal dimension was estimated as the slope of a fitted regression to log-log plot of, e.g. for Higuchi method $\langle L(k) \rangle$ versus k. Also we estimated two scaling regions corresponding to STV and LTV, respectively [33]. The separation (critical) time was 3s. In addition, in order to estimate both regions by one parameter, we also fitted the log-log plot with a second order polynomial which coefficients (first order and second order polynomial coefficient) correspond to the both STV and LTV. The **Approximate Entropy (ApEn)** is able to distinguish between low-dimensional deterministic systems, chaotic systems, stochastic and mixed systems [37]. ApEn(m,r) approximately equals the average of a natural logarithm of conditional probabilities that sequences of length m are close to each other, within a tolerance r, even if a new point is added. A slightly modified estimation of approximate entropy was proposed by [38] and resulted in **Sample Entropy (SampEn)**. This estimation overcame the shortcomings of the ApEn mainly because the self-matches were excluded. The used parameters for ApEn and SampEn estimation are: tolerance $r = (0.15; 0.2) \cdot SD$ (SD stands for standard deviation) and the embedding dimension $m = 2$ [39] The last of the nonlinear features was the **Lempel Ziv Complexity (LZC)** [40]. This method examines reoccurring patterns contained in the time series irrespective of time. A periodic signal has the same reoccurring patterns and low complexity while in random signals individual patterns are rarely repeated and signal complexity is high.

3.3   Feature Selection

Usually in most pattern recognition applications the feature extraction stage is followed by a feature selection stage [41] which reduces the input dimensionality, because in real world applications we tend to extract more features than necessary in an effort to include all possible information. However, sometimes some of the extracted features can be correlated, hence redundant information is likely to be included or sometimes some features are irrelevant to the application at hand and may negatively affect the performance of the classifier. The term "performance" refers to the training time required during the construction of the classification model or, which is the more serious side-effect, the discriminative capability of the classifier. In feature selection, a search problem of finding a subset of l features from a given set of d features, $l < d$ has to be solved in order to optimize a specific evaluation measure, i.e the performance of the classifier. There are a number of approaches that try to tackle this problem which can roughly be divided into three categories: filters, wrappers and embedded methods [42]. The filter approach ranks features based on a performance evaluation metric calculated directly from the data; the wrapper approach employs a predictive model and uses its output to determine the quality of the selected features and the embedded approach integrates the selection of features in model building. In this work a hybrid approach combining a filter and a wrapper approach was combined. More specifically RELevance In Estimating Features (RELIEF) was employed to rank the features and then based on the ranking the number of retained features was determined by directly estimating their performance using a predictive model. In the rest of the section we briefly present RELIEF whereas the wrapped stage is explained in more detail in Section 4.

RELIEF is a popular feature selection algorithm based on a weight vector over all features which is updated according to the sample points presented (the higher the weight the better the feature). The algorithm for a binary classification problem can be summarized as follows

---

Algorithm 1. RELIEF algorithm

Input: a data set $D = < \mathbf{x}_1, y_1, \ldots, \mathbf{x_M}, y_M >$, with $\mathbf{x}_i \in \mathbb{R}^N$ and $y_i \in \{-1, 1\}$
    for $i = 1, \ldots, M$
a relevancy cut-off (threshold) τ
a number of iteration T
begin
    i) initialize the weight vector to zero $\mathbf{w} = (0, 0, \ldots, 0)$
    ii) **for** $t \in T$ **do**
        pick at random an example $\mathbf{x}$
        for $i \in N$ **do**
            update the elements of the weight vector
            $w_i = w_i + (x_i - nearmiss(\mathbf{x})_i)^2 - (x_i - nearhit(\mathbf{x})_i)^2$
            where $nearmiss(\mathbf{x})$ and $nearhit(\mathbf{x})$ denote the nearest point to $\mathbf{x}$ in
            D that belong to the other and the same class, respectively.
        end
        iii) select the feature set whose members exceed the given relevancy
        cut-off (threshold) τ, $S = \{i | w_i > \tau\}$
    end
end

In our case the step $iii)$ was not involved. Instead we selected the highest 40 out of the total 54 features and then we employed a wrapper approach using the simplest form of search procedure, the "Best Individual" [43], in order to select the number of retained features. In other words after using RELIEF to rank the features we tested 40 different subsets starting from a subset containing the feature with the highest rank and we continued adding one feature at a time (the second best, the third best etc.) and we estimated their classification performance. The subset with the highest performance was determining the number of features involved in the estimation performance phase as it is will be presented in more detail in the next Section 4.

4   Classification Procedure

As it was pointed out in section 2, one class, the abnormal one, is heavily under-sampled in comparison to the normal one. This creates an extra challenge to the already difficult task of fetus well-being diagnosis. The class imbalance is a fundamental problem, arising when pattern recognition methods are dealing with real life problems, and many approaches have been proposed to overcome this situation [44]. In order to compensate for this imbalance we employed a popular technique which operates on the minority class creating artificial data, the Synthetic Minority Oversampling TEchnique (SMOTE). SMOTE is based on real data belonging to the minority class and it operates in the feature space rather than the data space [45]. The algorithm for each instance (in feature space) of the minority class introduces a synthetic example along any/all of the lines joining that particular instance with its k nearest neighbors that belong to the minority class. Usually after SMOTE the training set has approximately equal numbers of the 2 classes. However in this study our preliminary results suggested that more synthetic data from the minority class were needed. Therefore we selected to oversample the minority class by a factor of 18 using 27 (k=27) neighbors without trying to further optimize/tune the parameter settings of SMOTE. For testing the classification performance by making use of as many of the abnormal instances as possible we applied a 44 fold (stratified) cross validation procedure with each fold containing 1 abnormal instance and 12 or 11 normal instances. Therefore each time 43 abnormal instances were used for training and 496 or 497 normal instances and 1 abnormal instance and 12 (11) normal instances were saved for testing. During each fold we applied SMOTE to the abnormal instances, with the aforementioned parameters, while the normal instances remained intact. After the application of SMOTE an "inner" loop involved for the selection of the "optimal" number of features. During every fold RELIEF used all the training data (not the synthetic ones) to rank the features and then an inner loop was executed 4 times during which the data was randomly divided into training and testing (70/30) and a classifier was tested using 1 to 40 features (starting with the best feature and adding one feature at a time based on its ranking). Based on the average classification accuracy over these four repetitions the "optimal" number of features was selected. After selecting the number of retained features, the whole training set (with the inclusion of the data coming from the SMOTE stage) was used to train a classifier to be tested on the reserved testing set. In this work we employed the simplest member of the nearest prototype classifier family, the nearest mean prototype classifier, which assigns an instance to the class whose mean vector is closest to, during the inner loop procedure, and after that (after the selection of the number of features to retain) we employed the same classifier but within the adaboost framework in order to come up with a more powerful classification scheme. Adaboost which comes from adaptive boosting was first introduced by Freund and Schapire [46] is a general method for improving the performance of a week learner. It is the most well-known model guided instance selection for building ensemble classifiers. The basic steps of the algorithm are summarized as follows (following mainly the notation provided in [47]).

Algorithm 2. Adaboost algorithm

Input: a data set $D = < \mathbf{x}_1, y_1, \ldots, \mathbf{x}_M, y_M >$, with $\mathbf{x}_i \in \mathbb{R}^N$ and $y_i \in \{-1, 1\}$
      for $i = 1, \ldots, M$
k_{max} – the maximum number of weak learners to be included in the ensemble
C – a weak learner
begin
    i) initialize the weight vector $\mathbf{W}_1 = (1/M, 1/M, \ldots, 1/M)$
    ii) **for** $k = 1, \ldots, k_{max}$ **do**
        train weak learner C_k sampling D according to $\mathbf{W}_k$
        $e_k \leftarrow \sum_{i:C_k(x_i) \neq y_i} \mathbf{W}_k(i)$, where $C_k(\mathbf{x})$ is the output of the weak
        classifier for instance $\mathbf{x}$
        $\alpha_k \leftarrow \frac{1}{2} \ln \left(\frac{1-e_k}{e_k} \right)$
        $\mathbf{W}_{k+1}(i) \leftarrow \frac{\mathbf{W}_k(i)}{Z_k} \times \begin{cases} e^{-\alpha_k}, \text{ if } C_k(\mathbf{x_i}) = y_i \\ e^{-\alpha_k}, \text{ if } C_k(\mathbf{x_i}) \neq y_i \end{cases}$, Z_k is normalizing constant
    end
    iii) classify any new instance $\mathbf{x}$ using $G(\mathbf{x}) = sign \left(\sum_{k=1}^{k_{max}} \alpha_k C_k(\mathbf{x}) \right)$
end

In this work 40 nearest mean classifiers were employed. Trying to avoid any bias regarding the selection of the normal instance in each fold we repeated the procedure 5 times each time randomly reshuffling the normal instances creating an "outer" loop. By the outer/inner scheme we decouple the parameter selection stage from the estimation of the performance in an attempt to avoid getting optimistic results. The overall procedure is depicted in Fig. 1. The results of the 5 times repetition of the stratified 44-fold cross-validation procedure are summarized in the aggregated/cumulative confusion matrix, see Tab. 2.

As it can be observed with the specific setting we managed to have a balanced performance for both the normal and abnormal case. Regarding the feature selection process, Fig. 3 shows the number of times each feature configuration (number of features) has been selected over the 5x44 trials. As it can be seen usually a number between 10 and 20 was the most frequent configuration.

Regarding the ranking of the features by the RELIEF algorithm, Fig. 4a depicts the average ranking of the features (lower values are better) whereas Fig. 4b shows the number of times each individual feature was ranked.

Table 3 summarizes the top 10 features in terms of their average rank and Tab. 4 summarizes the top 10 features in terms of occurrences within the top 20 spot list.

Table 2. Cumulative confusion matrix of the proposed approach

		Predicted	
		Abnormal	Normal
Actual	Abnormal	141	79
	Normal	884	1656

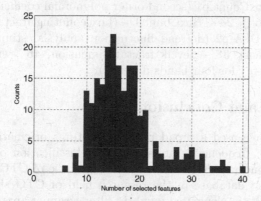

Fig. 3. Histogram of the number of selected features, through the "inner" loop procedure

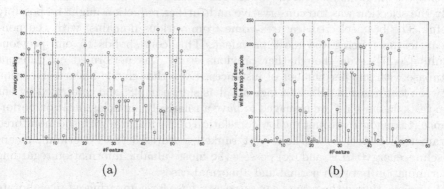

(a)   (b)

Fig. 4. Features ranking. a) Average ranking of individual features, b) Number of occurrences of individual features within the top-20 ranked list.

Table 3. The top 10 features selected by the RELIEF algorithm

Feature	7	13	44	17	36	38	24	26	47	46
Average rank	1.21	1.84	3.60	4.96	9.17	9.27	11.40	11.81	13.55	13.80

The numbers in the Table. 3 stands for the following features: **7** – STV-DeHann, **13** – meanBaseline (mean of FHR baseline), **44** – energy04_LF (low frequency energy for four bands division), **17** – lzc (Lempel Ziv Complexity),

Table 4. The top 10 features in terms of occurrences within the top 20 list

Feature	7	13	44	17	36	26	24	37	38	46
# occurrences	220	220	220	220	215	211	206	197	195	193

36: DFA_p1 (detrend fluctuation analysis estimated second order polynomial coefficient), **26** – BoxCount_p1 (second order polynomial coefficient estimate by box counting method), **24** – BoxCount_Ds (box counting fractal dimension on short scale), **37** – DFA_p2 (detrend fluctuation analysis estimated first order polynomial coefficient), **38** – Sevcik fractal dimension, **46** – energy04_HF_LF (high frequency energy for four bands division).

5   Discussion and Conclusion

In this work we have used a broad range of features originating from different domains (time, frequency, state-space) for classification of CTG records into normal and abnormal classes. We used a database of CTG records, which is one of the largest database in the research field of CTG signal processing and classification. We implemented a hybrid filter-wrapper approach for feature selection were roughly 25% of features was filtered out using RELIEF algorithm and the rest were coupled with a nearest mean prototype classifier for further redcing the dimensionality of the input space. Our results indicate that the probably the selection was too conservative and further reduction might be possibly useful. The best selected features come from various domains, with the non-linear features being the most prevalent. This corresponds to our previous results [20,27], even though that previous study was performed on different database with smaller number of instances. Even though the linear correlation between features and pH is not high at all as shown in [48] and confirmed by our own tests, in our case, we managed to have a relatively high classification performance (taking into account the low correlation of the features with the monitored parameter). Especially the four best ranked features 3 (STV-DeHann, mean-Baseline, energy04_LF, and lzc) possess the most valuable information regarding discrimination between normal and abnormal cases.

As we mentioned the results can serve as a base for comparing more elaborate classification schemes involving differt feature selection schemes and different classifiers. In our future work we intend to use other ranking methods such as the minimal-redundancy-maximal-relevance (mRMR) criterion [49], which can cope better with the problem of redundant information and we also intend to try replace the wrapper approach and use a random forest (RF) [50] to act upon a reduced number of features since the results of this work suggest that around 20 features could be a reasonable set of features, reducing this way the computational burden of the wrapper approch by taking advantage of the relatively quick trainig time of the RF. Moreover other state of the art classifiers such as the SVMs and the deep belief neural networks will be tested and compared against the "base" results that were derived from this study.

Acknowledgments. This work was supported by the following research programs: research grant No.NT11124-6/2010 of the Ministry of Health Care of the Czech Republic, MSMT-11646/2012-36 of the Ministry of Education, Youth and Sports of the Czech Republic, and CVUT grant SGS13/203/OHK3/3T/13.

References

1. Alfirevic, Z., Devane, D., Gyte, G.M.L.: Continuous cardiotocography (CTG) as a form of electronic fetal monitoring (EFM) for fetal assessment during labour. Cochrane Database Syst. Rev. 3(3), CD006066 (2006)
2. Bernardes, J., Costa-Pereira, A., de Campos, D.A., van Geijn, H.P., Pereira-Leite, L.: Evaluation of interobserver agreement of cardiotocograms. Int. J. Gynaecol. Obstet. 57(1), 33–37 (1997)
3. Blix, E., Sviggum, O., Koss, K.S., Oian, P.: Inter-observer variation in assessment of 845 labour admission tests: comparison between midwives and obstetricians in the clinical setting and two experts. BJOG 110(1), 1–5 (2003)
4. Chen, H.Y., Chauhan, S.P., Ananth, C.V., Vintzileos, A.M., Abuhamad, A.Z.: Electronic fetal heart rate monitoring and its relationship to neonatal and infant mortality in the United States. Am. J. Obstet. Gynecol. 204(6), 491.e1–491.e10 (2011)
5. Norén, H., Amer-Wåhlin, I., Hagberg, H., Herbst, A., Kjellmer, I., Maršál, K., Olofsson, P., Rosén, K.G.: Fetal electrocardiography in labor and neonatal outcome: data from the Swedish randomized controlled trial on intrapartum fetal monitoring. Am. J. Obstet. Gynecol. 188(1), 183–192 (2003)
6. Amer-Wåhlin, I., Maršál, K.: ST analysis of fetal electrocardiography in labor. Seminars in Fetal and Neonatal Medicine 16(1), 29–35 (2011)
7. FIGO: Guidelines for the Use of Fetal Monitoring. International Journal of Gynecology & Obstetrics 25, 159–167 (1986)
8. ACOG: American College of Obstetricians and Gynecologists Practice Bulletin No. 106: Intrapartum fetal heart rate monitoring: nomenclature, interpretation, and general management principles. Obstet. Gynecol. 114(1), 192–202 (2009)
9. Blackwell, S.C., Grobman, W.A., Antoniewicz, L., Hutchinson, M., Gyamfi Bannerman, C.: Interobserver and intraobserver reliability of the NICHD 3-Tier Fetal Heart Rate Interpretation System. Am. J. Obstet. Gynecol. 205(4), 378.e1–378.e5 (2011)
10. de Campos, D.A., Ugwumadu, A., Banfield, P., Lynch, P., Amin, P., Horwell, D., Costa, A., Santos, C., Bernardes, J., Rosen, K.: A randomised clinical trial of intrapartum fetal monitoring with computer analysis and alerts versus previously available monitoring. BMC Pregnancy Childbirth 10, 71 (2010)
11. Dawes, G.S., Visser, G.H., Goodman, J.D., Redman, C.W.: Numerical analysis of the human fetal heart rate: the quality of ultrasound records. Am. J. Obstet. Gynecol. 141(1), 43–52 (1981)
12. de Campos, D.A., Sousa, P., Costa, A., Bernardes, J.: Omniview-SisPorto 3.5 - A central fetal monitoring station with online alerts based on computerized cardiotocogram+ST event analysis. Journal of Perinatal Medicine 36(3), 260–264 (2008)
13. Task-Force: Heart rate variability. Standards of measurement, physiological interpretation, and clinical use. Task Force of the European Society of Cardiology and the North American Society of Pacing and Electrophysiology. Eur. Heart J. 17(3), 354–381 (March 1996)

14. Magenes, G., Signorini, M.G., Arduini, D.: Classification of cardiotocographic records by neural networks. In: Proc. IEEE-INNS-ENNS International Joint Conference on Neural Networks, IJCNN 2000, vol. 3, pp. 637–641 (2000)

15. Gonçalves, H., Rocha, A.P., de Campos, D.A., Bernardes, J.: Linear and nonlinear fetal heart rate analysis of normal and acidemic fetuses in the minutes preceding delivery. Med. Biol. Eng. Comput. 44(10), 847–855 (2006)

16. Van Laar, J., Porath, M., Peters, C., Oei, S.: Spectral analysis of fetal heart rate variability for fetal surveillance: Review of the literature. Acta Obstetricia et Gynecologica Scandinavica 87(3), 300–306 (2008)

17. Georgoulas, G., Stylios, C.D., Groumpos, P.P.: Feature Extraction and Classification of Fetal Heart Rate Using Wavelet Analysis and Support Vector Machines. International Journal on Artificial Intelligence Tools 15, 411–432 (2005)

18. Ferrario, M., Signorini, M.G., Magenes, G., Cerutti, S.: Comparison of entropy-based regularity estimators: application to the fetal heart rate signal for the identification of fetal distress. IEEE Trans. Biomed. Eng. 53(1), 119–125 (2006)

19. Gonçalves, H., Bernardes, J., Rocha, A.P., de Campos, D.A.: Linear and nonlinear analysis of heart rate patterns associated with fetal behavioral states in the antepartum period. Early Hum. Dev. 83(9), 585–591 (2007)

20. Spilka, J., Chudáček, V., Koucký, M., Lhotská, L., Huptych, M., Janků, P., Georgoulas, G., Stylios, C.: Using nonlinear features for fetal heart rate classification. Biomedical Signal Processing and Control 7(4), 350–357 (2012)

21. Georgoulas, G., Stylios, C.D., Groumpos, P.P.: Predicting the risk of metabolic acidosis for newborns based on fetal heart rate signal classification using support vector machines. IEEE Trans. Biomed. Eng. 53(5), 875–884 (2006)

22. Czabanski, R., Jezewski, M., Wrobel, J., Jezewski, J., Horoba, K.: Predicting the risk of low-fetal birth weight from cardiotocographic signals using ANBLIR system with deterministic annealing and epsilon-insensitive learning. IEEE Trans. Inf. Technol. Biomed. 14(4), 1062–1074 (2010)

23. Georgieva, A., Payne, S.J., Moulden, M., Redman, C.W.G.: Artificial neural networks applied to fetal monitoring in labour. Neural Computing and Applications 22(1), 85–93 (2013)

24. Georgoulas, G., Gavrilis, D., Tsoulos, I.G., Stylios, C.D., Bernardes, J., Groumpos, P.P.: Novel approach for fetal heart rate classification introducing grammatical evolution. Biomedical Signal Processing and Control 2, 69–79 (2007)

25. Sheiner, E., Hadar, A., Hallak, M., Katz, M., Mazor, M., Shoham-Vardi, I.: Clinical significance of fetal heart rate tracings during the second stage of labor. Obstet. Gynecol. 97(5, pt. 1), 747–752 (2001)

26. Chudáček, V., Spilka, J., Burša, M., Janků, P., Hruban, L., Huptych, M., Lhotská, L.: Open access intrapartum CTG database: Stepping stone towards generalization of technical findings on CTG signals. PLoS ONE (manuscript submitted for publication, 2013)

27. Chudáček, V., Spilka, J., Lhotská, L., Janků, P., Koucký, M., Huptych, M., Burša, M.: Assessment of features for automatic CTG analysis based on expert annotation. In: Conf. Proc. IEEE Eng. Med. Biol. Soc. 2011, pp. 6051–6054 (2011)

28. Cesarelli, M., Romano, M., Bifulco, P.: Comparison of short term variability indexes in cardiotocographic foetal monitoring. Comput. Biol. Med. 39(2), 106–118 (2009)

29. de Haan, J., van Bemmel, J., Versteeg, B., Veth, A., Stolte, L., Janssens, J., Eskes, T.: Quantitative evaluation of fetal heart rate patterns. I. Processing methods. European Journal of Obstetrics and Gynecology and Reproductive Biology 1(3), 95–102 (1971), cited By (since 1996) 13

30. Yeh, S.Y., Forsythe, A., Hon, E.H.: Quantification of fetal heart beat-to-beat interval differences. Obstet. Gynecol. 41(3), 355–363 (1973)
31. Pardey, J., Moulden, M., Redman, C.W.G.: A computer system for the numerical analysis of nonstress tests. Am. J. Obstet. Gynecol. 186(5), 1095–1103 (2002)
32. Signorini, M.G., Magenes, G., Cerutti, S., Arduini, D.: Linear and nonlinear parameters for the analysis of fetal heart rate signal from cardiotocographic recordings. IEEE Trans. Biomed. Eng. 50(3), 365–374 (2003)
33. Higuchi, T.: Approach to an irregular time series on the basis of the fractal theory. Phys. D 31(2), 277–283 (1988)
34. Kinsner, W.: Batch and real-time computation of a fractal dimension based on variance of a time series. Technical report, Department of Electrical & Computer Engineering, University of Manitoba, Winnipeg, Canada (1994)
35. Sevcik, C.: A Procedure to Estimate the Fractal Dimension of Waveforms. Complexity International 5 (1998)
36. Peng, C.K., Havlin, S., Stanley, H.E., Goldberger, A.L.: Quantification of scaling exponents and crossover phenomena in nonstationary heartbeat time series. Chaos 5(1), 82–87 (1995)
37. Pincus, S.: Approximate entropy (ApEn) as a complexity measure. Chaos 5 (1), 110–117 (1995)
38. Richman, J.S., Moorman, J.R.: Physiological time-series analysis using approximate entropy and sample entropy. Am. J. Physiol. Heart Circ. Physiol. 278(6), H2039–H2049 (2000)
39. Pincus, S.M., Viscarello, R.R.: Approximate entropy: a regularity measure for fetal heart rate analysis. Obstet. Gynecol. 79(2), 249–255 (1992)
40. Lempel, A., Ziv, J.: On the complexity of finite sequences. IEEE Transactions on Information Theory IT-22(1), 75–81 (1976)
41. Theodoridis, S., Koutroumbas, K.: Pattern recognition, 4th edn. (2009)
42. Guyon, I., Gunn, S., Nikravesh, M., Zadeh, L.A.: Feature extraction: foundations and applications, vol. 207. Springer (2006)
43. Webb, A.R.: Statistical pattern recognition. Wiley (2003)
44. Chawla, N.V., Japkowicz, N., Kotcz, A.: Editorial: special issue on learning from imbalanced data sets. ACM SIGKDD Explorations Newsletter 6(1), 1–6 (2004)
45. Chawla, N.V., Bowyer, K.W., Hall, L.O., Kegelmeyer, W.P.: SMOTE: Synthetic Minority Over-sampling Technique. Journal of Artificial Intelligence Research 16, 321–357 (2002)
46. Freund, Y., Schapire, R.E.: Experiments with a New Boosting Algorithm. In: International Conference on Machine Learning, pp. 148–156 (1996)
47. Duda, R.O., Hart, P.E., Stork, D.G.: Pattern classification, p. 1. John Wiley Section 10, New York (2001)
48. Fulcher, B., Georgieva, A., Redman, C., Jones, N.: Highly comparative fetal heart rate analysis. In: 2012 Annual International Conference of the IEEE Engineering in Medicine and Biology Society (EMBC), August 28-September 1, pp. 3135–3138 (2012)
49. Peng, H., Long, F., Ding, C.: Feature selection based on mutual information criteria of max-dependency, max-relevance, and min-redundancy. IEEE Transactions on Pattern Analysis and Machine Intelligence 27(8), 1226–1238 (2005)
50. Breiman, L.: Random Forests. Machine Learning 45(1), 5–32 (2001)

Maternal and Neonatal Healthcare Information System: Development of an Obstetric Electronic Health Record and Healthcare Indicators Dashboard

Juliano Gaspar[1,2,3], Junea Chagas[1], Gabriel C. Osanan[2],
Ricardo Cruz-Correia[3], and Zilma Silveira N. Reis[1,2,3]

[1] CINS – Centro de Informática em Saúde da Faculdade de Medicina da Universidade
Federal de Minas Gerais, Brasil
[2] Departamento de Ginecologia e Obstetrícia da Faculdade de Medicina da Universidade
Federal de Minas Gerais, Brasil
[3] CINTESIS – Centro de Investigação em Tecnologias e Sistemas de Informação em Saúde,
Faculdade de Medicina da Universidade do Porto, Porto, Portugal
{jgaspar,rcorreia}@med.up.pt,
{juneachagas,zilma.medicina}@gmail.com, osanangc@yahoo.com

Abstract. Development and use of specific information systems are nowadays a priority to healthcare organizations. This interdisciplinary paper, between Medicine and Computer Science, comprises an exploratory survey of obstetric and neonatal assistance during the process of birth by focusing on clinical records. It also involves the development of a healthcare indicators dashboard and the implementation of an Obstetric Electronic Health Record. All these efforts aim to improve quality of healthcare assistance. This prototype has been installed within a Brazilian university hospital. Preliminary results included 907 deliveries (from August to December 2012). The Cesarean section rate was 36.4% (65.4% high-risk pregnancy). Among the perinatal indicators, the Apgar Score bellow seven in the 5th minute was 5.56% and 0.93% whenever considering only fetuses defined as compatible with life. Disclosure and discussion of these data proved to be able to contribute to researches and to management of an obstetric healthcare unit.

Keywords: Healthcare Indicators, Electronic Health Record, Obstetric Data.

1 Introduction

1.1 The Current Challenges to the Maternal and Neonatal Care in Brazil

The new sanitary reality of Brazil consists of considerable progress in the mother & child care, especially on the latest decades [1]. Nevertheless, immense challenges need to be overcome to provide safe birth conditions [2], as there is still an undesirable rate of neonatal and maternal deaths in Brazil [3].

The latest decades have been marked by great scientific advances in the fields of maternal and perinatal health. As a result, it is not acceptable that the reproductive

M. Bursa, S. Khuri, and M.E. Renda (Eds.): ITBAM 2013, LNCS 8060, pp. 62–76, 2013.

process can result in maternal or fetal damages or even death. It was estimated that in 2008 there were three million births per year in Brazil, from which 1.2% died during the neonatal period [4]. Another relevant point is the fact that neonatal mortality became the main component in child mortality since the 90's and, unlike that observed in post-neonatal mortality, it has remained stabilized in high levels [5]. Most of neonatal deaths occur in the early neonatal period (0-6 days of life). From these 40% occur on the first day of life, but an expressive number of deaths occur during the first hours of life. This scenario reflects the strict relationship between infant mortality and prenatal and birth care in maternity hospitals [5]. Most of these deaths are considered preventable through timely access to qualified health services [6, 7].

Another major health care challenge is to reduce cesarean section and maternal mortality in Brazil [8]. In 2008, about 69 pregnant women died per 100.000 live births in Brazil due to pregnancy related causes [4]. About 65% of maternal deaths occur during delivery [9]. The proportion of women giving birth in hospitals has been as high as 95% in Brazil, which by itself does not guarantee the quality of birth care [10]. Incidence of cesarean section in Brazil has been around 50% over the last years [4]. Nevertheless, its progressive growth in the last decades was not followed by decrease on the indicators of poor maternal and neonatal outcome [11].

Efforts are being made to enhance the knowledge about the conditions or weaknesses in the health care system that can contribute to a poor maternal and neonatal outcome. In this context, computerized information about pregnancy and birth are important for evaluating the quality of maternal and neonatal care. Clinical records not only constitute vital documentation to the service but also constitute a set of data potentially able to identify local needs by means of using electronic systematization procedures [12]. If there are reliable tools to evaluate, continuously, the prenatal, the delivery and neonatal care, it allows the healthcare managers to monitor the quality and the impact of their actions to improve health assistance [13].

1.2   The Importance of the Maternal and Child Health Indicators to the Public Management of Healthcare and Improvement of the Care Process

Given the complexity of the concept of health, the task of measuring it by using health indicators becomes equally complex. Considered as a whole, those indicators should reflect the health status of a given population and should serve to the surveillance of health conditions [10].

Due to the exponential growth of computerization in hospitals over the past years, the volume of electronic health records (EHR) has increased substantially. Additionally, there is interest in the analysis of such data to support clinical and hospital managerial decisions in the healthcare organizations [14, 15].

Provision of reliable information derived from solid data is vital to support professionals and technicians in decision making and to the support of hospital managers. In this context, information systems are responsible for generating, analyzing and spreading such data [16].

In order to achieve the projected goals, the choice of a set of indicators to monitor a given reality will depend on the specific objectives. In such cases, the computerization of the records and the integration between the health information systems (HIS) become determinant steps to make it possible to compose population indicators based on clinical data. A sample of this trend for more specialized indicators is the group of those that aim at benchmarking situations of maternal near-miss (it accounts near death conditions). This indicator is being extensively used to try to determine conditions that are life-threatening for the mother. It regards the compilation of information, that which characterize the adverse maternal result based on a pre-established set of severe situations, the complexity of which comprises the benchmarking and proper recording of clinical history details [17, 18].

Another matter in the choice and the analysis of indicators is the various contexts of the healthcare. The indicators must be analyzed under the light of situations, conditions and local rules so that they can be used in a standardized and universal form in order to allow comparability between the different health services, localities and even between different countries. For instance, a very much used indicator of neonatal results is the Apgar score in the 5th minute of life of the newborn. This score is useful as a parameter to evaluate neonatal hypoxia, a condition that can indicate inadequate birth assistance. However the Apgar score may suffer a negative influence by fetal malformations, extreme prematurity or even a use of some drugs by the mother. In this way in those countries such as Brazil, where the abortion of malformed is not allowed, a low Apgar score may be related to a fetal condition (malformation) and not the perinatal hypoxia.

On the other hand, the quality of the health indicators also depends on the quality of the data, the pool of cases, the population risk, the precision and integration of the HIS comprehending the recording, collecting and transmitting the data. The availability of reliable and valid data is an essential condition to support decision making [19].

1.3   The Importance of the Computerization of Clinical Data and Health Attendance Processes

The use of healthcare information systems (HIS) is now a real need for the care services and for great research groups worldwide. It consists of a fundamental strategy to obtain care approaches with better scientific quality, for it is a tool capable of guaranteeing the accumulation of standardized data in a continuous and safe way throughout time and simultaneously between different locations, as long as it may guarantee the integration of data. The possibility of interconnecting healthcare data generated in different sites also represents an advance in the scientific research for health/illness state determinants. Besides the analysis of the distribution of diseases throughout time, it allows the understanding, prediction, casual evidencing, prevention and impact evaluation of interventions in health care [20].

The search for a greater quality on data and an adequate EHR system also constitute an opportunity for revaluation of the health care itself. The failure of the implementation of a data system however commences in the concept phase due to the lack of effective participation of the future system users and developers causing the criteria efficiency on the field, the usability (ease of handling) and the interoperability (ability of integrating to other systems) to be poorly defined and understood by the participants [12]. Besides this care, it is noteworthy that the initial investment in the standardization aiming the exchange of information (interoperation) between different healthcare electronic systems, designed in distinct contexts, often for specific purposes, results in profit in the form of acknowledged social and economic benefits.

Hence, the fields of information technology and informatics in health are assuming an ever more relevant role in the care of people's health. Brazil has shared the global concern about the advances in health policies based on broad range Information Systems of quality, yet guided by the improvement of healthcare actions and services. In line with the international context, the Brazilian Ministry of Health established as one of its main priorities the definition of a national policy for information and health informatics, understanding it as essential to the achievement of broader equity, quality and transparency on the healthcare services intended by the National Health System (NHS). The 12th National Health Conference of 2003 ruled in favor of the implementation of articulated information and communication policies, outlined so as to reinforce its democratization in every respect.

1.4   Aim

This paper aims to propose, implement and evaluate an Obstetric Electronic Health Record and to develop a healthcare indicators dashboard. It also aims to evaluate the hypothesis that interdisciplinary work between medicine and computer science is essential to develop an effective obstetrical and neonatal software.

2   Methods

It comprises applied research, interdisciplinary scientific study between Medicine, Information and Computer Science. The study takes place in the scenario of a public maternity hospital of tertiary reference (high-risk pregnancy) bound to a public Brazilian university. Clinical data recorded on paper documents on all obstetric hospitalizations were used as object of study. The compilation of the data for the automated calculation of the performance indicators and its transcription to a standard electronic form was performed by medical personnel from August to December 2012. Simultaneously and integrated, the team of medical informatics develops software specific to the automated calculation of the performance indicators (both individual and collective) of this health service.

Due to the present need for developing technical capabilities to the construction and maintenance of the system, multidisciplinary professionals were involved, such as doctors, system analysts and statisticians, by means of a partnership between Federal University of Minas Gerais (UFMG) and University of Porto (UPorto).

The present study was approved by the Committee for Research Ethics of Federal University of Minas Gerais under the registration number (Brazil Plataform: 10286913.3.0000.5149) all human research principles respected. The proposition was presented and supported by Hospital das Clínicas of the UFMG and Health Informatics Center of the Faculty of Medicine of the UFMG.

The methodology of the study was divided into three distinct stages:

- Exploratory survey of the scenario of obstetric and neonatal care based on the compilation of data recorded in paper clinical documentation, obtained from the parturient and neonate healthcare;
- Simultaneous development of an Obstetric EHR;

- A survey study of the indicators of quality of obstetric and neonatal healthcare, to result in the development of an electronic tool to calculate automatically a new predetermined set of indicators of maternal and neonatal health.

To carry on such stages the following steps were defined:

1. **Variables definition:** Definition of the clinical and administrative variables needed to compose a condensed version of an obstetric EHR;
2. **Transcription of values from manuscript to computer:** Reading and collection of the paper data recorded to transcribe into a specific electronic form, as the service does not yet feature an integrated electronic record system and most of the data on delivery is registered in manuscripts. Therefore, there was a predefined set of data considered vital to generate automatically the intended indicators. Among them: the pregnant profile, the indicators for cesarean section and the maternal and neonatal outcomes of every admission on the defined period;
3. **Identification and definition of maternal and neonatal health indicators:** Setting more adequate indicators to benchmark the quality of obstetric and neonatal care in maternity hospitals, as well as the corresponding formulas and calculations for each indicator.
4. **HIS implementation:** Implementation of specific software for input/collection, storage and analysis of obstetric data. A software capable of composing the obstetric EHR and of automatically generating the indicator panels on maternal and neonatal heath;
5. **Prototype installation for tests:** Installation of a prototype to be tested in the maternity hospital and to be integrated to the other electronic management systems existing in that health facility;
6. **Evaluation of first results:** Evaluation of the results and the impact of the health indicators over the quality of health care attendance;
7. **Creation and release of reports:** Both individual and collective, for internal and external disclosure of the indicators of birth care in this public healthcare service, making it possible to improve the quality of care management, performed transparently and shared with its authors.

3   Results

3.1   Variables Definition

The database developed has 126 variables that can be grouped in three groups according to the nature: epidemiological, clinical and administrative. The epidemiological data constitute a socially defined set of data capable of identifying individual characteristics of the pregnant women with potential to determine damage to health, such as: age, home city, parity, prenatal care original site and gestational risk factors. The clinic data refer to the care and the practices employed during the delivery. Those include from the admission scores (gestational age, characterization of labor, use of the partogram) to the data on the type of delivery (vaginal or surgical, pharmacological or non-pharmacological analgesia), episiotomy, birth conditions (alive/dead, weight, Apgar), evolution during the stay in the hospital (complications/ICU) and the conditions for the maternal and neonatal discharge. Administrative data are time and date of admission, birth and discharge, so as of the hospital account processing to be paid by NHS.

3.2   Transcription of Values from Manuscript to Computer

The first stage of this study involved the reading and collection of epidemiological, clinical and administrative data from the obstetric and neonatal clinical records from admissions conducted in the Maternity Otto Cirne from Hospital das Clínicas of Federal University of Minas Gerais (Brazil). All the admissions during the period of August to December 2012 were considered, totalizing 1094 paper clinical records. In a second stage, these collected data were inserted into an Obstetric EHR developed for this project.

3.3   Identification and Definition of Maternal and Neonatal Health Indicators

In this stage of the development of the project, the implementation of the indicators contracted with the government or with the quality goals aimed by the management of the maternity hospital was prioritized. These indicators were grouped in modules that offer the user, through a panel of graphic information, a quick perception of the quality of the attendance made by the maternity. Among the indicator modules worth mentioning are: maternity prodution indicators, governament contracted indicators, deliveries indicators: vaginal deliveries and cesarean sections, perinatal indicators, care quality indicators by professional profile.

The indicators dashboard feature now 70 distinct indicators that approach epidemiological, clinical, administrative and qualitative aspects, defined as essential to evaluate the quality of the care provided. In this paper will be analyzed the first results presented by the following indicators:

- **Cesarean section rate:** Measures the proportion regarding cesarean section deliveries. It is calculated by dividing the number of deliveries by the total of deliveries in a given period.
- **High-risk patients rate:** The proportion of high-risk patients is calculated by dividing the number of patients with high-risk gestation by the total patients admitted in the period. These were considered gestations of high-risk protocol by the Brazilian Ministry of Health.
- **Length of stay:** This indicator is obtained by summing the number of admission days of all the patients and dividing by the total of patients. The admission period of each patient may be obtained by subtracting the discharge date of a given patient from the respective admission date.
- **Apgar score after the 5ᵗʰ minute below 7:** The incidence of low Apgar score of 5th minute was calculated by dividing the number of neonates with Apgar score below 7, after the 5th minute of life, for the total of neonates born alive in the period.
- **Apgar score after the 5ᵗʰ minute below 7 (adjusted):** A correction to the Apgar score was made necessary given the expressive number of malformed or unviable neonates in this reference maternity. Therefore, the occurrence of

revised low Apgar score after the 5th minute was calculated considering only the viable neonates, that is, excluding those with severe malformations (pulmonary hypoplasia, polimalformed, anencephalic) or extremely premature (gestational ages between 22 and 25 weeks).

- **Breastfeeding in the 1st hour of life rate:** This indicator is one of the predictors of the humanized delivery care. It is obtained by calculating the proportion of living newborns fed in their first hour of life from the total of born alive.

- **Breastfeeding in the 1st hour of life rate (adjusted):** As to the Apgar score indicator, some correction was also demanded by this indicator. The revised rate of newborns fed in the 1st hour is obtained by excluding the inviable neonates and those for whom the nursing is contraindicated (seropositive HIV and Hepatitis B, for instance).

- **Maternal discharge status of patient (DSP):** Four outcomes to the hospital discharge have been pre-set to the women: hospital discharge, hospital transfer to another hospital, exit against medical advice and maternal death. These four possibilities were scaled and presented in the indicator panels, by calculating the proportion of each category in relation to the total of admitted women.

- **Neonatal discharge status of patient (DSP):** Accordingly, established to maternal outcomes for each newborn and agreed to among the health staff are seven possible conditions: hospital discharge, hospital transfer to another hospital, exit against medical advice, neonate retained, death previous to admission, death during admission and neonatal death. For each possibility was calculated the frequency in relation to the total of births, all featuring the indicator panel.

3.4   HIS Implementation

In the development of this prototype was chosen the interactive and incremental model in order to obtain preliminary results quickly. Therefore, from the initial prototype it is possible to evaluate results, determine new enhancements and features, reducing the risks of failure in the accomplishment of the proposed goals.

To the development of the implementation of the prototype were used OpenSource technologies as PHP, HTML and JavaScript languages and the database was built with MySQL, choices made to reduce development and final product costs to the public health network.

This project originated software product called SisMater[1] (Maternal and Neonatal Health Information System). The SisMater features two main modules: the Obstetric Electronic Health Record (Fig. 1) and the module Obstetric and Neonatal Indicators, which includes reports of both individual performance of the health professionals and of the general obstetric health service (Fig. 2).

---

[1] SisMater: Sistema de Informação em Saúde Materna e Neonatal (in portuguese).

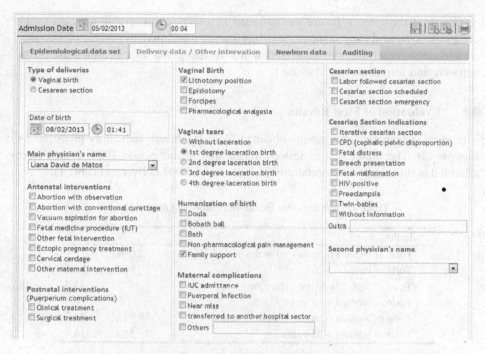

Fig. 1. Obstetric Electronic Health Record

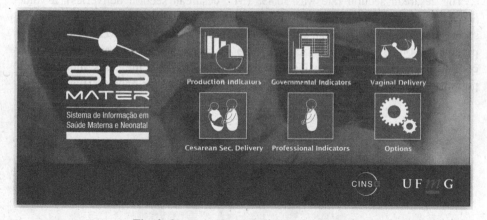

Fig. 2. Obstetric Health Indicators Dashboard

3.5 Prototype Installation for Tests

From a paper form pre-defined, that served to collect retrospective data from the manuscript clinical record forms the data was inserted in the module of Obstetric EHR of SisMater. Then a test version was installed locally in a single computer in the maternity hospital. Only a few participants and helpers of the project were granted access to input data.

In a third stage, the software was installed in a web server, granting access to the electronic system to any computer existing in the maternity, by means of previous authorization. After that, the obstetrics internship was given lessons on the use of the software and were encouraged to use it.

3.6   Evaluation of First Results

This section features some results obtained from the health indicators produced by this project. The production indicator informs that in this period 1094 women were admitted in the maternity hospital, from which derived 907 deliveries (Table 1).

Table 1. Dashboard for Health Indicators of Obstetric Care

Indicators	N	%
Total of women hospitalized in the period	1.094	-
Total of births	907	82.9
Total of high risk women attended	478	43.7
Total of women < 15 years	9	0.8
Total of women between 15 and 35 years	922	84.3
Total of women > 35 years	163	14.9

The cesarean section rate was 36.4% (312 cesarean section deliveries), a value above the goal set to the hospital in 30% (Fig. 3). The Obstetric Quality indicator shows that among the cesarean sections made in the period 65.4% were conducted on pregnant women with risk factor to the occurrence of maternal or neonatal complications (Fig. 4), what is justified by the complexity of the cases attended by a public maternity hospital that is tertiary reference (high-risk pregnancy) (Fig. 4).

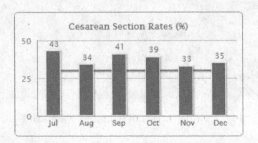

Fig. 3. Cesarean section rate per moth

A reduction of cesarean section taxes of 3.6 percentile points is noticed when compared the three first months of the project (39.3%) to the three last months analyzed (35.7%).

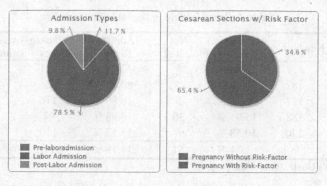

Fig. 4. Admission Types and Cesarean Sections in Pregnant women with Risk

The average admission periods, according to the delivery type were presented in Table 2. All of them were below the target, set by the hospital managers.

Table 2. Length of stay (LOS) in the maternity service, considering modalities of birth

Indicators	Time	Target
Global LOS	1 day and 19:12h	-
General Births LOS	2 days and 0:00 h	
Vaginal deliveries LOS	1 day and 9:36 h	1 day
Cesarean sections LOS	3 days and 2:24 h	2 days
Ante partum hospitalization LOS	2 days and 7:12 h	

The Apgar score < 7 in the 5th minute (gross), grouped in the first 5 months analyzed was 5.56% while the same rate considering only fetuses compatible with life (adjusted) was 0.93% (Table 3).

Table 3. Rate of 5-minute Apgar Score below 7

Months	Apgar Score (gross rate)	Apgar Score (adjusted rate)
August	1.69 %	0.56 %
September	9.39 %	1.10 %
October	6.25 %	1.70 %
November	5.56 %	1.23 %
December	4.79 %	0.0 %
Average	5.56 %	0.93 %

Government's Goal: Rate below 1.5%

Table 4 features analysis of the rates of the indicator that contains the breastfeeding in the 1st hour of life of the newborn, in both gross and adjusted rates. It is important to highlight that a significant amount of the clinical records did not bring this information filled in, was noticed that between the 894 born alive, 62.2% of the clinical records did not contain this information recorded. Even when considering only the clinical records that contained this data and adjusting the calculation only in the cases it was pointed out, the average rate of breastfeeding in the 1st hour of life was 41.32%, below that established by the government (100%).

Table 4. Breastfeeding in the 1st hour of life

Months	Newborn	Gross Rate	Not Indicated	Adjusted Rate	Missing Values	Adjusted Rate (without missing)
August	177	11.86 %	5	12.20 %	121	41.17 %
September	182	10.43 %	5	10.73 %	106	26.76 %
October	177	16.38 %	5	16.86 %	116	51.78 %
November	188	15.95 %	30	18.98 %	94	46.87 %
December	170	10.58 %	6	15.12 %	119	40.00 %
Average	178.8	13.04 %	51	14.78 %	111.2	41.32 %

Government's Goal: 100 % of the births

Regarding the maternal DSP, it was observed that 99.5% of women admitted in the maternity hospital had hospital discharge. Despite the efforts of the health professionals the maternal death rate was 0.1%, corresponding to one maternal death in the analyzed period (Fig. 5).

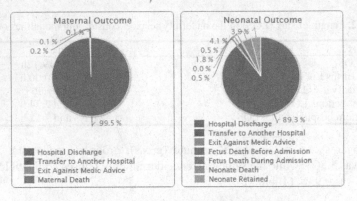

Fig. 5. Maternal and Neonatal DSP

Regarding the neonatal DSP the rate of hospital discharge was 89.3%, 3.9% of which were still admitted (retained) by the end of this research period. The occurrence of neonatal death was 4.1%. The fetal death rate previous to the admission was 1.8%. The occurrence of fetal death during admission was 0.5%. Transfer of neonates to other hospitals occurred in 0.5% of the cases (Fig. 5).

3.7   Creation and Release of Reports

The first reports containing individual performance results of the period August to December 2012 by health professionals, as well as for the whole obstetric and neonatal health care service were released. Those indicators were shared and discussed among the medical staff, managers and professors in charge of the training of the resident doctors from the university hospital. Such release of the first results took place in different meetings with either manager of the maternity or medical teams. The external audition teams representing the government control over the provision of public services by the National Health Service (NHS) also received and discussed preliminarily the first reports.

The release of the individual performance of the medic professionals allowed for the first time in the university hospital to retrieve a detailed description of their instructional activities in the teaching maternity hospital. It brought to individualized attention the quantitative values of the obstetric procedures made under its responsibility, as much as the corresponding maternal and neonatal results. It also allowed the managers of the public health service to obtain specific data on relevant aspects of the maternity services, declared in its services contract. The more accurate identification of the profile of the attendance of this service allowed making considerations on the possible causes of the current performance, in face of some indicators demanded by the government, questioning the goals themselves.

4   Discussion

The computerization of hospital clinical records of birth is innovative subject to the obstetrics and is promising at the same time due to the relevance of its potential in causing positive impact on the healthcare assistance challenges in that area. However, to make this kind of approach viable it is necessary to bring together the clinic-obstetrical experience in following the deliveries on one hand, especially high-risk where the complications are more frequent, and on the other hand, an experienced service that stands out in the scientific and technological production of health information services.

The reports generated by the prototype to the maternity proved it useful far beyond describing automatically and for the first time the indicators of results in this healthcare service, it allowed the positioning of the institution in face of the goals preset by the hospital and by the government. Another relevant gain was the questioning of the goals themselves. By not considering the characteristics of a university reference service, whose profile of admission of maternal and neonatal high-risk pregnancies is beyond 50%, were proposed some unreachable goals. The care challenges are not the same faced by general maternity hospitals that aid women with low gestational risk and therefore are not comparable to each other, under the same goals to be reached.

The preview of the indicators in a graphic format, throughout the five months, allowed specialists and connoisseurs of the maternity hospital's reality important perceptions, just by analyzing visually the evolution of the rate of cesarean sections. Among those, the verification that cesarean sections were frequent between pregnant women associated with some risk factor. Important discussions followed the release of the first reports and motivated the search for a new procedure standard by the health professionals.

The real-time preview of the administrative indicators by the managers of this healthcare service, as much as the admission lengths, creates formerly non existing opportunities of adopting measures more quickly and with better potential of resoluteness to challenges such as: the occupation of high-risk beds by public care reference services.

The proposition of an adjusted rate to the frequency of the Apgar scores in the 5th minute under 7, between neonates compatible with survival by the birth was one of the most important products of the discussion on the performance indicators automatically generated by the SisMater panels. This was possible thanks to the setting of a clearer clinical-epidemiological profile of the pregnancies attended in this health facility, bringing to light the frequency and severity of the illnesses that complicate pregnancies in the admitted cases, in the studied period. The percentile of severe malformed fetuses was estimated, easing the understanding of the specific needs of this healthcare facility and its indicators particularly distinct from those of other public maternity hospitals, out of the universities. It is important to comment that in Brazil the anticipated interruption of pregnancies due to malformed fetuses is not legally possible, except for the recently established case of anencephalic fetuses.

Similarly, the rate of breastfeeding in the 1st hour of life, adjusted for the cases in which it is really recommended, brought up the discussion on the expressive number of pregnancies complicated by AIDS/HIV positive, as much as for the quality of the clinical records and the procedures on birth existing on paper forms. The estimated values, even if adjusted, of this humanizing practice (Table 4), are much below the projected by the government (goal 100%). They don't reflect well the reality of this practice routinely applied by this maternity's health professionals. Such finding raised attention upon the responsibility of the professionals responsible for the annotations in the paper forms used in this health facility today.

Due to referring to an experience of computerization of health clinical records in a tertiary public maternity inside a school hospital, the study intends to contribute to the establishment of new patterns of reference to the indicators of performance, which may come to benchmark more reliably the quality of the care provided in health services of this nature. In that order, it will also involve a study on the proposition of new adjusted rates, when made necessary, intended to be more adequate to the specific particularities of fetal medicine and the care of high-risk pregnancies.

5   Conclusion

The presentation of results in the graphic form allowed a simple and more direct reading of the most relevant values, being, doubtlessly an added value in the assistance to the management of the control of efficiency of the health facilities.

SisMater has shown, already in the presentation of the first results, efficient in the proposition of indicator panels of maternal-infantile and neonatal quality of the maternity hospital. Despite that the service doesn't feature an Electronic Health Record System (HER-S) it was possible to obtain knowledge of the reality of the maternal and neonatal care. From this comprehension was possible to identify some deficiencies of the service and set new action priorities, in order to reach, in the future, better results in the quality of the care offered to the parturients and neonates.

Last but not least, stands out that knowing in detail the reality of the studied population allowed to establish new adjusted indicators and, therefore, contest/debate, either in the scope of the hospital or in the governmental sphere determined goals impossible to attend in facilities of high-risk pregnancy attendance.

5.1   Future Works

During the development of this project and the analysis of the preliminary results, other stages were defined. New functionalities were foreseen from the perception of the assistance challenges, formerly hidden or not understood due to the lack of automated information, such as:

- Integration to the clinical records generating in the neonatal unity, that shall involve specific data regarding the neonate in three modes of admission: the neonatal ICU, the rooming-in and the kangaroo mother care;
- Conducting training on the use of the system (SisMater) for the resident doctors of neonatal pediatrics and the insertion of this new group of users;
- Enhancement of the automated panels of individual and collective performance indicators of obstetric and neonatal care, from the analysis and discussion of the present indicators;
- Integration between the system and the preexisting software of the maternity;
- Operationalization of the normalization of the data, adapting them to the international standards of interoperability with the Open-HER defaults, adopted by the Brazilian National Health bureau.

Acknowledgements. The researchers of this project acknowledge the support given by the CAPES (Coordenação de Aperfeiçoamento de Pessoal de Nível Superior) and FAPEMIG (Fundação de Amparo a Pesquisa de Minas Gerais) of the development of this project.

References

1. Victora, C.G.: Maternal and child health in Brazil: progress and challenges. Lancet 377(9780), 1863–1876 (2011) (Epub May 9, 2011)
2. do Leal, M.C., Moura da Silva, A.A., Dias, M.A., Nogueira da Gama, S.G., Rattner, D., Moreira, M.E., et al.: Birth in Brazil: national survey into labour and birth. Reproductive Health 9(1), 15 (2012) (Epub August 24, 2012)
3. Fonseca, S.C., Coutinho, E.D.S.F.: Pesquisa sobre mortalidade perinatal no Brasil: revisão da metodologia e dos resultados. Cadernos de Saúde Pública 20, S7–S19 (2004)
4. MS-Brazil. Ministério da Saúde do Brasil. Sistema de Informática do SUS. Brasília (2012)
5. MS-Brasil. Manual dos comitês de mortalidade materna. In: Saúde, à S.d.A. (ed.), 3a edn., Brasília, p. 104 (2007)
6. Reis, Z.S.N., Waleska, R.S.: Sofrimento Fetal Agudo. Perinatologia Básica, 3rd edn., pp. 182–184. Guanabara, Rio de Janeiro (2006)
7. Reis, Z.S.N., Osanan, G.C.: Sofrimento Fetal Agudo. Manual de Ginecologia e Obstetrícia, 5a edn., p. 344. COOPMED (2012)
8. WHO. Caesarean section without medical indications is associated with an increased risk of adverse short-term maternal outcomes: The 2004-2008 WHO Global Survey on Maternal and Perinatal Health. BMC Med. 8, 71 (2010)
9. MS-Brasil, Saúde Md, Estratégicas DdAP. Política Nacional de Atenção Integral à Saúde da Mulher Princípios e Diretrizes. In: Saúde M. (ed.) Brasília: Ministério da Saúde, p. 82 (2007)

10. MS-Brasil. Indicadores basicos para a saude no Brasil: conceitos e aplicacoes. In: Ridipa, S. (ed.) 2 edn., p. 349. Organização Pan-Americana da Saúde, Brasília (2008)
11. Reis, Z.S.N., Pereira, A.C., Correia, R.J.C., Freitas, J.A.S., Cabral, A.C.V., Bernardes, J.: Análise de indicadores da saúde materno-infantil: paralelos entre Portugal e Brasil. Revista Brasileira de Ginecologia e Obstetrícia 33, 234–239 (2011)
12. Shortliffe, E.H.: Biomedical Informatics: Computer Applications in Health Care and Biomedicine, 3rd edn., 1060 p. Springer Science, New York (2006)
13. Cabral, A.C.V.: Fundamentos e prática em obstetrícia, 1st edn. Atheneu Editora, São Paulo (2009)
14. Silva-Costa, T., Marques, B., Freitas, A.: Problemas de Qualidade de Dados em Bases de Dados de Internamentos Hospitalares. In: 5a Conferência Ibérica de Sistemas e Tecnologias de Informação, Santiago de Compostela (2010)
15. Freitas, A., Brazdil, P., Costa-Pereira, A.: Mining Hospital Databases for Management Support. In: IADIS Virtual Multi Conference on Computer Science and Information Systems, pp. 207–212 (2005)
16. Pinto, R.: Sistemas de informações hospitalares de Brasil, Espanha e Portugal - Semelhanças e diferenças, Mestrado, FIOCRUZ, Escola Nacional de Saúde Pública Sergio Arouca, Rio de Janeiro (2010)
17. Souza, J.P., Cecatti, J.G., Haddad, S.M., Parpinelli, M.A., Costa, M.L., Katz, L., et al.: The WHO Maternal Near-Miss Approach and the Maternal Severity Index Model (MSI): Tools for Assessing the Management of Severe Maternal Morbidity. PloS One 7(8), e44129 (2012) (Epub September 7, 2012)
18. Say, L., Pattinson, R.C., Gulmezoglu, A.M.: WHO systematic review of maternal morbidity and mortality: the prevalence of severe acute maternal morbidity (near miss). Reproductive Health 1(1), 3 (2004) (Epub September 11, 2004)
19. OPS. Indicadores de Salud: Elementos básicos para el análisis de la situación de salud. Boletín Epidemiológico 22(4), 5 (2001)
20. MS-Brasil, SUS. BMdSDdId. A construção da política nacional de informação e informática em saúde: proposta versão 2.0 (inclui deliberações da 12. Conferência Nacional de Saúde): Ministério da Saúde, Secretaria-Executiva, Departamento de Informática do SUS (2005)

Data Acquisition and Storage System in a Cardiac Electrophysiology Laboratory

Matěj Hrachovina[1,2], Michal Huptych[1], and Lenka Lhotská[1]

[1] BioDat Research Group, Department of Cybernetics, FEE CTU, Prague
{hrachmat,huptycm}@fel.cvut.cz, lhotska@labe.felk.cvut.cz
[2] Institute of Physiology, 1st Medical Faculty, Charles University, Prague

Abstract. The paper discusses an important issue emerging in electrophysiological experiments, namely collection, storage and classification of large volumes of heterogeneous data from multiple sites represented by multiple incompatible and non-standard measuring devices. The main focus is on the description of a particular setup used in an experimental laboratory of cardiac electrophysiology.

1 Introduction

Electrophysiology plays a very important role in ensuring accurate clinical diagnoses. Many diseases cause symptoms that manifest far from the injured or deceased tissues. Locating and treating all the affected areas of the body is essential for proper patient care. Electrophysiology allows for the investigation of abnormal electrical signals in the body tissues [1]. It provides quantitative data to clinicians, supporting diagnostic processes and evaluating the treatment success. Often, some biological parameters as those measured in electrophysiology are more useful in assessing symptom severity than existing clinical measurement scales. Their objective nature removes subjective assignment of scores to symptom severity, subsequently leading to better informed health care decisions. Before new diagnostic approaches and treatment procedures are applied to humans they are tested on experimental animals. During these experiments large volumes of data from various devices are collected. In this paper we describe such experimental setting and the most important issues connected with data acquisition and storage.

2 Motivation

Cardiac electrophysiology experiments on living organisms are a means of designing new approaches to curing various heart related issues, such as arrhythmia, heart attack, left ventricle insufficiency and others. In order to assess the benefits of new methods proposed, relevant and reliable data need to be collected in a standardized manner. Only noise-free and correctly time-stamped, synchronized and labeled datasets can be used to draw relevant conclusions about the effects

M. Bursa, S. Khuri, and M.E. Renda (Eds.): ITBAM 2013, LNCS 8060, pp. 77–87, 2013.
© Springer-Verlag Berlin Heidelberg 2013

of various actions performed during an experiment. After all, when an experiment is finished, the collected data is the only scientific proof of the effects of a new approach. Moreover, the data acquired from repeated experiments must be comparable.

3   Data Collection

There are three basic types of data which are being collected during the experiment: electrical and non-electrical biosignals originating from the living body and human inputs originating from the staff performing the experiment.

3.1   Data Types

Non-electrical variables describe the vital parameters of the organism and include: ventilation rate and volumes, blood pressure in different parts of the body, composition and saturation of blood gases, cardiac output, rates of medication served intravenously. The data is mostly collected through service outputs of commercial medical devices, such as bedside monitors, ventilators and others, although a workaround can be used in some devices, which provide a limited number of analog signal outputs. These devices are mainly designed to monitor current state, rather than collect and archive the measured signals, which leads to problems discussed later, such as difficulties in obtaining raw signals for future evidence. The electrical biosignals originate from the heart and other muscles, and can be recorded directly after amplification. This data is collected by devices designed for data collection, namely an intracardiac ECG (EGM) and an admittance pressure-volume system. Human inputs are mainly notes describing deviations from standard experimental protocol or other events, that cannot be recorded as signals. An example of such input can be a note about changing a catheter position or about serving some medication not monitored by intravenous pumps.

3.2   Devices and Parameters

In this section we describe the parameters and functions of devices used during electrophysiological experiments. Their list is shown in Table 1.

The bedside monitor is a universal device with multiple measuring ports. Modular measuring accessories can be attached to its standardized inputs, depending on actual needs. It can measure up to 5 different pressures, usually intravenously from different parts of the body. Measurement range for pressure depends on the sensors used and is usually $0 - 300$ mmHg with 1 mmHg resolution and an accuracy of ± 1 mmHg ± 1 digit. Its 12-lead ECG has input dynamic range ± 5 mV and CMRR[1] greater than 95 dB. Accuracy of SpO_2 measurement is ± 2 digits. Temperature is updated every 3 seconds and is measured with $\pm 0.1°C$ accuracy.

---

[1] Common mode rejection ratio, $CMRR = 10 \log_{10} \frac{A_d}{A_{cm}}^2$.

Table 1. List of devices used in the experimental laboratory

Device type	Measured variable	Units	Sampling frequency	Data output
Bedside monitor	invasive pressure	mmHg	unknown	Digital through HL7, Analog for synchronization
	ECG	mV		
	SpO$_2$	%		
	CO$_2$	mmHg		
Ventilator	Inspiratory volume	l	50 Hz	Digital through RS-232
	Expiratory volume	l		
	Flow	l/min		
Oxygenation monitor	SvO$_2$	%	unknown	Digital through RS-232, Analog for synchronization
	Cardiac output	l/min		
Intracardiac ECG	EGM	μV	3 kHz	Digital through FTP
Admittance pressure volume system	RV Pressure	V	400 Hz	Analog
	RV Volume	V		
Blood parameter monitoring system	Chemical composition	mmHg, ml/min, mmol/l, pH	0.16-55 Hz	Digital through RS-232
A/D converter	Analog signals	mV	400 Hz	Digital through USB

Accuracy of CO$_2$ measurement is ±4 mmHg [2]. As this is a commercial device from Nihon Kohden, sampling frequencies of individual signals are unknown as well as algorithms used to process them. We also don't know the resolution of the A/D converter, so the bedside monitor cannot be used to obtain true raw data in the technical sense. However for medical interpretation the processed data is good enough. The bedside monitor was not designed to record data. However it supports TCP/IP protocol and can be connected into a network. HL7 protocol is then used to outsource data to an external computer. A virtual HL7 Gateway server is set up on the external computer. It is used to communicate with the monitor and a client recording the data via HL7 query and answer messages [3]. The query message defines, which signals will be returned as a response. Maximum data capacity of the message is 64 KB, which implies that the length of a signal segment transmitted depends on the number of signals requested [4]. The maximum length of signal in one message is 20 seconds. The actual data is recorded by a client, which facilitates the communication with the server and translates the received messages into a standard data container format, in our case MFER [5]. The communication pattern is outlined in Figure 1. The MFER file format was designed for storing multiple channels of medical data with different sampling rates.

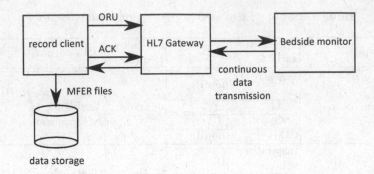

Fig. 1. Illustration of device↔client communication in HL7 protocol

The ventilator is primarily designed for breathing support, not for data collection. The data is recorded by connecting the device through service RS-232 port to a computer. A utility designed by the device manufacturer downloads waveforms of basic parameters with sampling frequency 50 Hz. It also records the device status and triggered alarms every second [6].

The oxygenation monitor measures the blood oxygen saturation and continuous cardiac output invasively by thermodilution method. As with other devices not designed for scientific purposes, the resolution and accuracy of the A/D converter used is not known, as well as the sampling frequency. Datasets from this monitor are collected from the service RS-232 port. The device sends all measured parameters every 2 seconds as plain ASCII text.

The intracardiac ECG is designed to record the time propagation of electrical potentials along the cardiac muscle. Signals are measured invasively by catheters with two to twenty electrodes. For reference purposes it also measures regular 12-lead ECG using surface electrodes. A significant benefit of using this device, compared to other devices, is that all filters can be switched off and true raw data can be recorded. Input range is ±37.6 mV with maximum ±1.5 mV DC offset and CMRR greater than 100 dB at 60 Hz. Sampling rate for all channels is 3 kHz and a 14 bit A/D converter is used for digitalization. The data is saved in binary format, one sample at a time for all channels, coded as 16 bit signed integer values. Although the recorded data is only meant to be reviewed on the device, there is a simple way to download it into a regular PC for further processing using FTP protocol.

The admittance pressure-volume system is the only diagnostic device used, which was designed solely for scientific purposes. It is used to measure pressure and volume of the left ventricle with special catheters using admittance volumetry. It is basically a pressure to voltage transducer and an amplifier with analog signal pre-processing. It only supports analog data output.

The blood parameter monitoring system allows continuous real-time measurement of eleven critical blood gas parameters using optical fluorescence and reflectance method. The parameters are measured with $\frac{1}{6}$ Hz to 55 Hz sampling frequency, the measurement ranges and accuracies can be found in [7]. The data are collected through a RS-232 connector, the device sends the measured parameters as plain text in ASCII format.

For devices that do not support digital output or for backup purposes, analog signals are collected by a measuring A/D converter. Parameters of the A/D converter are software determined. For most channels the input range is set to ± 10 V and sampling frequency 400 Hz. When digitalized to 16 bits the resolution is 313 μV, the quantization error is ± 2.5 LSB and CMRR is 100 dB. The A/D converter is connected through USB to a PC which saves the measured data.

3.3 Experiment Setup

Basic setup consists of a bedside monitor, ventilator, intracardiac ECG, blood oxygenation monitor and an event recorder. The bedside monitor is the basic instrument used by the medical staff to get information about the immediate effects, which the intervention is causing to the living body. In basic setup it measures ECG, arterial and venous pressure, SpO2 and CO2. These readings are used to adjust the doses of medicaments supplied, or to take other actions to keep the organism alive. The recorded data is later used to see what the condition of the organism was in different phases of the experiment. The living body needs to be kept under general anesthesia and therefore needs lung support. The ventilator measures aspiratory and expiratory volumes, pressures and respiratory rates as well as breathing gases composition. The intracardiac ECG measures electrical potentials from inside the heart through catheters in specific positions. Blood oxygenation monitor measures parameters indicating the state of heart and circulatory system, especially cardiac output and blood oxygen saturation. The event recorder serves as a human input device. It is basically a PC with a touchscreen and an application designed to record timestamped notes of deviances from the standard experiment protocol. It is used to record any event that is not represented by measurable data, which could have an effect on interpretation of the experiment outcome. An ilustration of signal and data paths is outlined in Figure 2.

3.4 Main Issues

The main problems that have to be solved to collect data of any future value are calibration, noise and synchronization. Apart from these three there are two more problems that need to be taken care of during the experiment design. First one is human error. It is very common that something is omitted during preparation for measurement or that an important setting is incorrect. To minimize this, a standard protocol for each experiment is outlined in the event recorder, where all experimental steps are marked as finished or pending. The other problem that can only be solved by experiment design is that the majority of commercial medical diagnostic devices are not intended for scientific purposes and are therefore unsuitable for data acquisition. These devices are built as black boxes and their technical parameters such as filter parameters, measurement accuracy, quantization accuracy are not provided by the manufactures, which degrades the

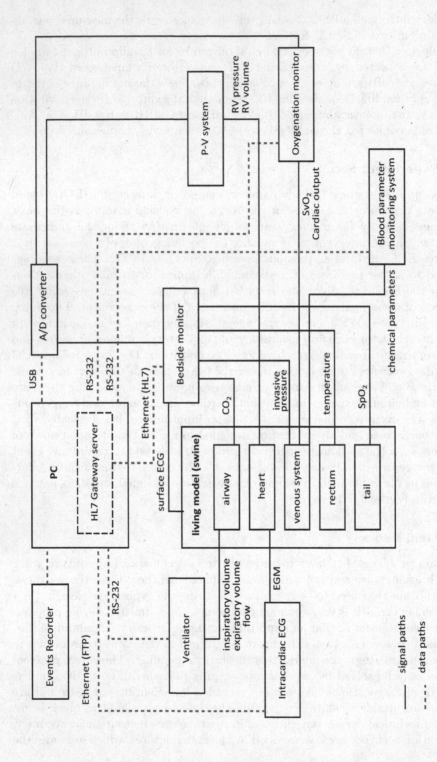

Fig. 2. Data paths

technical value of the measured data. They also lack standardized PC outputs, which makes the data collection system hard to manage and update.

Calibration. To obtain the data that could later be quantified, all signals need to be recorded with referenced levels. We achieve this by using calibrated sensors. The calibration applies to non-electrical signals, as voltages are measured directly during digitalization process. Chemical sensors need to be calibrated regularly using calibration gases. Pressure sensors cannot be calibrated, but their zero offset needs to be set before each experiment. In medical standards all pressures are measured with reference to atmospheric pressure, so that 0 mmHg equals the current pressure in the room.

Noise. Noise is an integral part of any electrical signal measurement. It can be filtered out of the signal by post-processing, but only to a certain degree, so it is very important to record the data with as low noise level as possible [8]. Commercial medical devices comply with CISPR11 and should have Electromagnetic Compatibility (EMC) certification [9] to ensure they do not affect each others operation and readings. But there are other devices in the laboratory that are not well shielded for electromagnetic (EMG) noise, for example the A/D converter in the pressure-volume system. Special care needs to be taken when positioning such devices, so that they are as far away from sources of EMG radiation as possible. Main known sources of noise in the mentioned setup are devices used for intervention, although some measurement apparatus may contribute too. The noise they produce affects the measurement of electrical biosignals, but in extreme situation it could lead to malfunction of the measuring devices. A reliable source of broadband EMC noise is electrocoagulation, which is used for tissue cutting. Another example is an electrical warming blanket, which we use to stabilize the body temperature in cold conditions. These two can ruin all electrical measurements, because the nature of the noise they produce make it impossible to filter out any relevant data. A common additive noise is 50 Hz sine wave from the mains. It comes from voltage transformers of various laboratory devices, but also from computers and displays. It is easily filtered out from the signal using comb filters. In the experimental environment where multiple channels with different signal amplitudes are collected, the wiring is very important. To minimize the noise induction, all electrode cables in one channel should be twisted together and their paths should be as straight as possible to minimize the EMG radiation on loops. The electrode cables used to measure weak signals (below 100 μV) should be properly shielded to reduce induction. All devices must have one common grounding and if possible, their power supply should be galvanically separated from the measurement circuits to reduce induction of the 50 Hz sine wave from mains into the measured signal. Other sources of noise are also known, such as cell phones and other wireless networks, but so far there has not been any significant effect observed in the experiment setup.

Synchronization. For later interpretation of different signals' correlation, time synchronization is important. There is a number of possible approaches.

The first approach is to timestamp each signal against a reliable time source, such as a NTP server. This can be managed easily with data that is outsourced from service connectors, where the timestamp can be added to each data block received. This is also the case with data collected from the bedside monitor using the HL7 protocol, which uses the gateway time in the header. This approach however is only valid provided the sampling period of the recorded signal is longer than the deviation tolerance of the time source, and the delay of the transition between the device and the PC adding the timestamp is negligible.

The second approach is to use a synchronization pulse, which could be observed in all measured signals. This is a bit tricky, as stimulation that would have immediate and measurable effect on both electrical and non-electrical parameters is hard to find. Defibrillation impulse is a possibility, but it has two drawbacks - it usually cannot be used before the end of the experiment and it may affect the operation of some devices, such as the pressure-volume system. Another problem of this approach is that a discontinuity in any measurement would lead to loss of synchronization. This can be overcome by periodically adding a sequence of synchronization impulses to the electrical signals. The sequence can vary in time, so the composition of the pulses can code for the time from the start of the experiment. An illustration of a possible pulse sequence is shown in Figure 3.

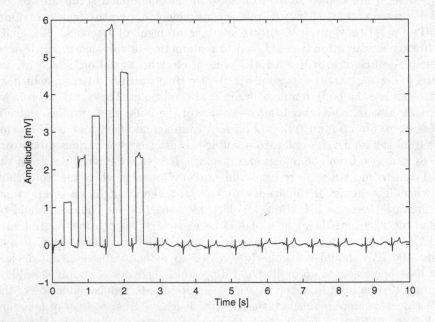

Fig. 3. Illustration of a synchronization pulse superposed on a ECG signal

Another approach can be used in long recordings. We have found that long-term trends of different parameters show similarities, so if one parameter is taken as a reference, the others can be synchronized according to it. This approach is however seldom used for we have not yet proved it scientifically and thus we cannot judge its accuracy and reliability.

The approach we use varies according to the planned experiment, to best fit the setup and protocol used, however time-stamping of each signal is used in every setup.

Adaptability. The experimental laboratory is a facility with technical equipment, which serves different medical teams to prove their hypotheses. That means that the data acquisition and storage system must be easily rearrangeable to suit different objectives from time to time. It also means it has to be ready to incorporate a new device if needed.

4   Data Storage

The experiments generate extensive amounts of data. The monitored time between preparation of the body and the end of the experiment takes 5 hours on average. Events resembling important clues to what lead to the outcome of the experiment can be anywhere in that period. That means all the data need to be recorded continuously and stored for future reference and evaluation. If the evaluation is to be effective, the data also needs to be properly labeled and sorted. The best way to do this is to store raw data in an ordinary file system and have a database with labels, references to raw data and information about the experiment [10].

4.1   Database Structure

The database is constructed from tables and relations according to logical resemblance with the experiment components, its general structure is shown in Figure 4. The basic block is the Event table, which holds records of individual events that occur during an experiment. The experiment itself is represented by the Record table. The relation between Record and Event is 1:n, because there are many events in an experiment, but each event can only belong to exactly one experiment. Each event can be of a different type, namely a Diagnosis, Intervention, Medication, Measurement or a combination. Having these in separate tables instead of a column in the Events table is beneficial in case of adding a new diagnosis or intervention. Each event can also be associated with the person that performed it, which is helpful in tracing responsibility for crucial decisions. Every event can have one or more measurements. Each measurement can be performed by only one Device, but one device can perform many measurements. The Measurement is defined by the Device therefore it has it as a foreign key. The Sensor table is not bound to the Device table as it is expected the Measurement can only be performed by one sensor from a device. One measurement can

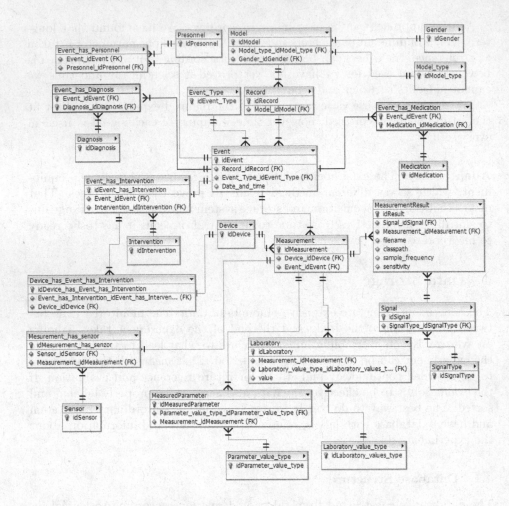

Fig. 4. ER Diagram of the database structure used

produce a Signal, which points to the location of the raw data in the file system. This way any Event can be paired with a corresponding section of a signal easily.

4.2 Design Limitations

In some relationships intersection tables have to be defined in order to reflect the exact type of the instance used. For example it would be ineffective defining all possible defibrillation energies in the Intervention table. Instead, the values would be added in the intersection table Event_has_intervention.

4.3 Future Amendments

When completely finished, the structure should be a unified system for data collection and automation. This implies developing an application that would display all the available signals for one instance with notes and other background information, which would aid in quick and synoptical evaluation of the recorded data. Another forthcoming change is programming an application, which would feed the measured signals into the database as blocks of samples in unified format rather than only saving a reference to file with raw data encoded according to the device it comes from.

5   Conclusion

Although we described a particular experimental setting in the paper, the idea and also the hardware and software setup is reusable for other types of experiments where large volumes of heterogeneous data are collected. Novelty of this setup is not in the individual techniques used, but in their combination into an automated system. The problem of collecting relevant datasets from cardiac electrophysiology experiments can be solved by obeying simple rules. This general proposal can be applied to any scale and type of experiment. When observing wiring paths and following set procedures, time synchronized and noise-free data can be obtained. It can also be easily sorted into a database which aids quick searching and sorting of desired data. Collected datasets can help in speeding up the evaluation of the investigated intervention methods. It can be used for automated analysis, which is the next focus of our research.

Acknowledgement. Research described in the paper has been supported by the CTU Grant SGS13/203/OHK3/3T/13.

References

1. Sigg, D.C.: Cardiac Electrophysiology Methods and Models, 1st edn. Springer (2010)
2. Kohden, N.: BSM-2301K Operator's Manual
3. Health Level Seven International (2012), http://www.hl7.org
4. Kohden, N.: HL7 Gateway Server QP-993PK Protocol Specification, 1st edn (2008)
5. MFER: Medical waveform description Format Encoding Rules Mfer, Part I, Version 1.01 (2003), http://www.mfer.org
6. Hamilton Medical: Hamilton S1 Operator's Manual, 624302/00 (October 2010)
7. TERUMO CVS: TERUMO CDI Blood Parameter Monitoring System 500 (2012), http://www.terumo-cvs.com/products/ProductDetail.aspx?groupId=1&familyID=47&country=1
8. Liang, H., Bronzino, J.D., Peterson, D.R.: Biosignal Processing: Principles and Practices. CRC Press (2012)
9. European Commission: Directive 2004/108/EC (2012), http://ec.europa.eu/enterprise/sectors/electrical/emc/index_en.htm
10. Huptych, M.: Multi-layer Data Model. PhD thesis, Czech Technical University in Prague (2013)

Proposing a Novel Monitoring and Early Notification System for Heart Diseases

Efrosini Sourla[1], Athanasios Tsakalidis[1], and Giannis Tzimas[2]

[1] Department of Computer Engineering & Informatics, School of Engineering,
University of Patras, Rio Campus, 26500, Patras, Greece
{sourla,tsak}@ceid.upatras.gr
[2] Department of Applied Informatics in Management & Economy, Faculty of Management and
Economics, Technological Educational Institute of Messolonghi, 30200, Messolonghi, Greece
tzimas@teimes.gr

Abstract. The spectacular penetration of Smartphones has introduced a new field of software applications' development. The constantly growing use of mobile applications for medicine, enables complete and systematic monitoring of chronic diseases for both health professionals and patients. In this work, an integrated system is presented for lifelong cardiologic patient monitoring, early detection and optimal management of emergency cases. The key features of the proposed system are: (a) management of emergency cases where the patient sends a chief complaint and the cardiologist evaluates the situation and proceeds to a series of actions, (b) management of patients' EMRs, (c) recording of patients' measurements of vital signs performed at home in regular basis and (d) interconnection to Microsoft HealthVault platform. Our system's contribution lies in: (a) optimal patient monitoring at home and early response in cases of emergency, (b) enhanced communication between collaborating parties and (c) maintenance of detailed medical data repositories, in common format, for scientific research.

Keywords: medical informatics, cardiology, health applications, mobile devices, e-health, telemedicine.

1 Introduction

Health (or Medical) Informatics has been defined as the study and implementation of structures to improve communication, understanding and management of medical information. The objective is the extraction, storage and manipulation of data and information and the development of tools and platforms that apply knowledge in the decision-making process. The advent of internet has broaden the scope of Medical Information Systems (MISs) and led to the development of distributed and interoperable information sources and services [10]. MISs are developed to serve every possible medical need and produce all kinds of medical information in various formats, including texts, numbers, pictures, static and dynamic images, etc. This heterogeneous information can then be integrated without the need of medical personnel.

M. Bursa, S. Khuri, and M.E. Renda (Eds.): ITBAM 2013, LNCS 8060, pp. 88–102, 2013.

Additionally, the spectacular penetration of mobile phones in the technological arena and their transformation into Smartphones, has introduced a new field of software applications' development. Smartphones have been employed widely in health care practice [1]. The level of their use is expected to increase, especially if they are enriched with doctor suitable functions and software applications. The lack of such applications is noticed even in countries with leadership role in mobile technologies, as it is mentioned in [4]. The impact of mobile handheld technology on hospital physicians' work practices and patient care is systematically reviewed in [5], where the authors recommend on future research about the impact of the mobility devices on work practices and outcomes.

In [14] five ways are presented of how mobile applications and smartphones will transform healthcare: (a) patients have improved access to care, since the requirement for them and doctors to be in the same location in the same time is eliminated in the digital age, (b) patients show improved engagement, since the enhanced services of mobile apps eliminate many existing deficiencies and difficulties – long lines, complexity, lack of transparency of cost and quality, (c) the explosion of inbound data from sensors and medical devices will create new opportunities for healthcare professionals and contribute to the accuracy of transferred data, (d) Medicare fraud will be reduced, since digital apps have the ability to track people and transactions in space and time and in the future will allow Medicare to correlate claims data with location and time and detect frauds in real-time, (e) digital apps will make health care safer by giving patients tools to manage their own health and have every medical information stored to or directly accessed from their smartphone.

An example which successfully combines MISs with the advantages and capabilities of Smartphones, in Orthopaedics, is the integrated system that was developed for recording, monitoring and studying patients with Open Tibia Fractures [3]. The authors participated in the development of the system, which is based on web and mobile applications. Primary goal was the creation of a system that contains most of the scientifically validated data elements, reducing this way omission and improving consistency, by standardizing the reporting language among medical doctors. The system's web and mobile interfaces are designed to require almost no text entry and editing and are based on the traditional medical way of acting, thus making it a doctor friendly system.

This paper presents an integrated system for lifelong cardiologic patient monitoring, early detection of emergency cases and optimal process management of the emergency incident. The proposed system consists of web applications, web services and smart phone applications. The system involves the participation of Cardiologic Patients, Cardiologists, General Practitioners, it allows interaction and instant communication between the corresponding parties, as well as Hospitals and out-Hospital health sectors.

The rest of the paper is organized as follows: Related work is presented in Section 2, along with our motivation. Section 3 presents a thorough description of the implemented system along with a first round evaluation. System's benefits are discussed in Section 4. Finally, Section 5 hosts conclusions and future work.

2   Related Work

2.1   Mobile Health Applications

Most health applications in online markets are native applications, patient oriented or medical doctor (MD) oriented. In most cases, the patient oriented health applications are exploited only by patients and the information gathered is not available directly to physicians, through a communication channel. Moreover, the MD oriented health applications serve specific purposes, mostly for educational and quick access to medical literature reasons. On the other hand, mobile applications that are part of medical research projects, frequently store information and send it to collaborative servers for additional processing and disposal to physicians.

The use of individual mobile health applications that have been developed to serve specific purposes, is widely spread. The need for such applications is apparent in every major online market for mobile applications including Android Market, Apple Store, Samsung Apps, etc. Applications developed for Cardiology record blood pressure and cardiac pulses, applications for Diabetes record blood glucose [2], for Obesity, they record calories and diet [6], for Dementia they use GPS to monitor the patient [7-8] and applications for chronic diseases target mobile phones with sensors and detect tachycardia or respiratory infections [1].

2.2   Medical Applications for Cardiology

Many mobile applications for cardiology have been developed in order to enhance medical doctors' and medical students' research experience [12], [15], [16] such as: (a) applications that present a 3D prototype of a human heart and allow users to observe the heart from any angle, (b) calculators with commonly used formulas in cardiovascular medicine, (c) electrocardiography (ECG) guides with samples of different types of ECG, (d) guideline tools for clinical practice and diagnosis and (e) decision support tools including several criteria and cases. All the above applications are addressed to medical staff, mainly for educational reasons and quick access to literature data, useful for Medical Doctors. However, they are not suitable for Cardiologic Patients.

A web environment for monitoring cardiologic patients is Heart360 Cardiovascular Wellness Center [11], sponsored by the American Heart Association and American Stroke Association. Heart360 allows patients to monitor their blood pressure, blood glucose, cholesterol, weight, nutrition and physical activity, while receiving education and information specific to their condition. Heart360 utilizes Microsoft HealthVault [13]. More specifically, patients are able to: (a) collect and record their blood pressure, blood glucose, cholesterol, weight, nutrition and physical activity habits, (b) set goals and track their progress, (c) view their data in charts and graphs that they can print out and share with others involved in their family health, (d) manage multiple user accounts and (e) get news and articles of potential interest based on their store of health information. This application is patient oriented and does not offer any substantial help to Cardiologists, or General Practitioners.

2.3   Our Motivation

The advent of new technologies has strongly benefited Healthcare Systems. Nevertheless, the new technologies incorporated have to be safe, reliable, user-friendly and offer proven solutions, since they are directly applied to humans, in particular patients and MDs.

Our system is motivated from the need to use the advantages of new technologies in the field of clinical medicine and home healthcare. Moreover, another important motive is the detected deficiencies in the existing medical systems and applications (web or mobile). Cardiology is a first line emergency medical specialty that has to deal not only with a variety of chronic diseases, but also with emergency incidents. Therefore, if the proposed system is proved to be an effective tool in the hands of Cardiologists, General Practitioners (GPs) and Cardiology Patients (CPs), then the same methodology can be used for the development of other Medical Specialty systems.

Moreover, to the best of our knowledge, till now there is not any system available that incorporates the following characteristics: (a) deals with emergency situations in an automated way, (b) is both patient and medical doctor oriented, (c) provides access to end-users through multiple channels (web, mobile), (d) utilizes existing and state-of-the-art medical platforms, such as Microsoft HealthVault and (e) is easily expandable and deployable. Thus, the detected deficiencies and the willingness to offer enhanced services to patients and medical doctors, motivated us to develop a system that incorporates these characteristics into an integrated functionality.

The proposed system is part of an ongoing project, which accrued from the collaboration between the Computer Engineering and Informatics Department and School of Medicine of the University of Patras and the General Hospital of Patras "Agios Andreas" and is currently in testing phase with the following involved parties: a) non-Hospital and Hospital Cardiologists, b) non-Hospital and Hospital General Practitioners and c) Cardiologic Patients. A demo version of the software is publicly available at: http://www.biodata.gr/cardiosmart365/

3   Description and Services

3.1   Introduction

The work presented here is an integrated system for lifelong cardiologic patient monitoring, early detection of emergency cases and optimal process management of the emergency incident. The system design and implementation use the well known service oriented architecture (SOA) to maximize interoperability and scalability, as well as user interface design techniques for optimal presentation. The system consists of web applications, native mobile applications for Smartphones and loosely coupled web services. The system involves the participation of Cardiologic Patients, Cardiologists, General Practitioners, allows interaction and instant communication between the corresponding parties, as well as Hospitals and out-Hospital health sectors. The three main services the system offers are (Fig. 1):

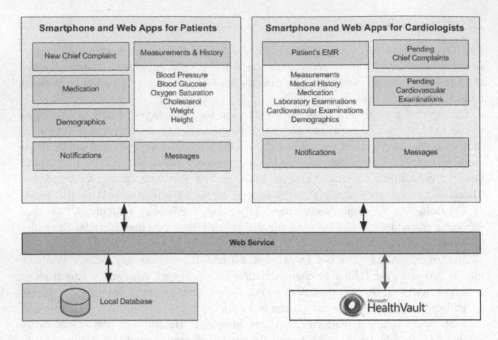

Fig. 1. System's Architecture: Offered Services

- Early detection and management of emergency cases where the patient sends a chief complaint (CC) and the cardiologist evaluates the situation and proceeds to a series of actions.
- Management of cardiologic patients' complete electronic medical records (EMRs). EMRs are managed by cardiologists responsible for each patient.
- Recording and management of patient's measurements of vital signs performed at home in regular basis, such as: blood pressure, blood glucose, oxygen saturation, weight and height.

Moreover, the system provides an integrated message management module, for optimal communication between end users and instant notifications. The module also includes automated messages.

The system implements a client - server architecture. Authorized end-users have access to the integrated system through client applications, a web application and a native mobile application for smart phones, with friendly and easy to use interfaces. Great emphasis has been given in the design of user – friendly and functional interfaces for both physicians and patients. In particular, the interface of mobile devices is designed in such a way to require the minimum volume of typing data.

In order to achieve platform independence, the client applications communicate and exchange data with the database through web services, which allow data interchange through heterogeneous systems. The web services provide functionality with which specific information can be accessed by client applications after authenticated access.

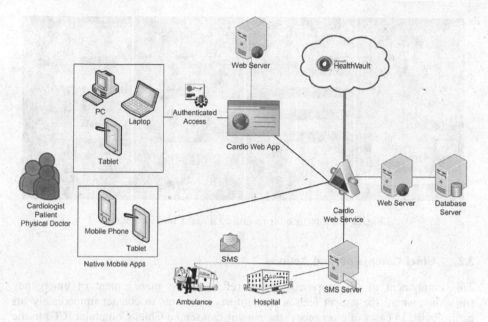

Fig. 2. System's Architecture: Components Interconnection

The proposed system utilizes the Microsoft HealthVault platform as a back-end platform, to store and manage important information of patients' EMRs and measurements, into a uniform format. Microsoft HealthVault [13] is a back-end cloud-based platform, based on EMR systems, which provides a privacy and security enhanced foundation that can be used to store and transfer information between a variety of e-healthcare customer's applications (desktop, web and mobile ones), hospital applications and healthcare devices. It also offers tools to solution providers, device manufacturers and developers, in order to build innovative new health and wellness management solutions.

The confidentiality and security of the patient's medical information is ensured by the authentication and encryption scheme of our system and the security mechanisms of HealthVault [13]. Only authorized users have access to the integrated system, after providing the right credentials. Moreover, the information exchanged between the web services and the web or mobile applications is strongly encrypted. Last but not least, HealthVault provides access to medical data from third party applications only when the latter are aligned with security certificates created and uploaded to Microsoft HealthVault by the HealthVault Application Manager. Although a third party application can obtain a certificate, it won't have access to a specific patient's medical information unless the patient himself gives explicit permission through his Health-Vault account from the HealthVault web site. This methodology ensures that sensitive user data, such as username and password, do not need to be transferred to the third party applications [9]. The system's architecture is shown in Fig. 2.

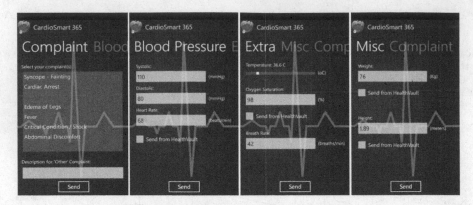

Fig. 3. The interface for recording a new Chief Complaint

3.2   Chief Complaints and Actions

This component of the proposed system refers to the management of emergency situations, where the patient feels a discomfort and wants to contact immediately his cardiologist. In cases of emergency, the patient can send a Chief Complaint (CC) to the

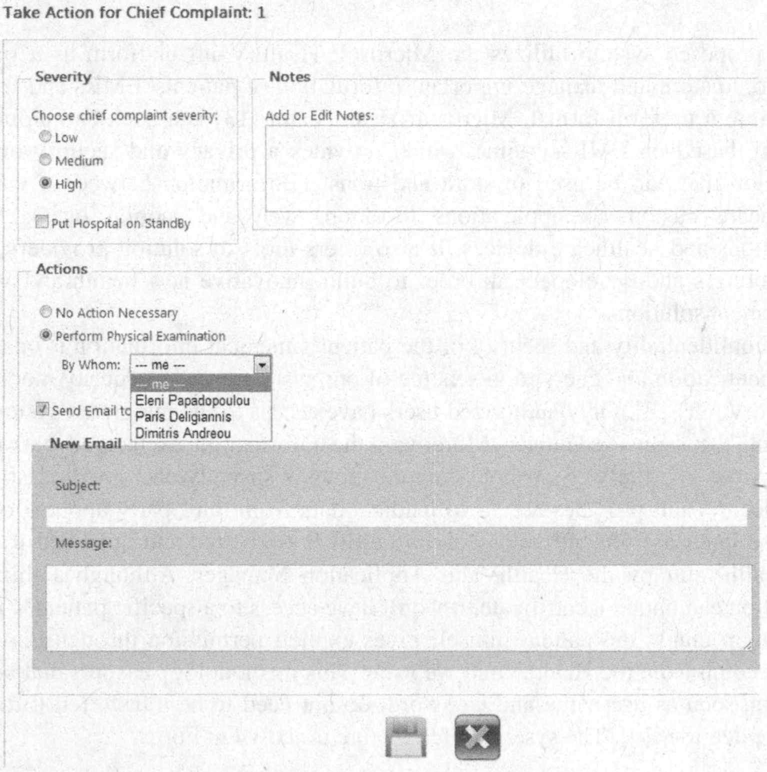

Fig. 4. Taking Actions for Chief Complaint

cardiologist, through a web interface or a native application for mobile devices (Fig. 3). The CC is a subjective statement made by a patient describing the most significant or serious symptoms or signs of illness or dysfunction that caused him/her to seek health care. Patient checks one or more of the most common chief complaints: chest discomfort, dyspnea, syncope – fainting, cardiac arrest, palpitations, edema of legs, fever, critical condition/shock, abdominal discomfort, or describes what he/she feels. Optionally, the patient enters the values of current measurements for blood pressure, temperature, oxygen saturation, breath rate, weight and height, or orders the system to send the most recent ones stored in HealthVault. When the new CC has been registered, the corresponding cardiologist is informed by an automated message (email, SMS, internal message) for the new emergency.

The cardiologist evaluates the situation and proceeds to a series of actions, depending on the severity of the incident. Through the system's web interface (Fig. 4) or mobile application, the cardiologist characterizes the severity of the CC as low, medium or high and optionally adds some notes. Subsequently, he/she checks if a cardiovascular examination must be performed and selects the examiner who will perform it (himself or a nearby available general doctor). In this case, an automated message is send to everyone concerned: cardiologist, patient and potential general doctor. If the situation is of great severity, the cardiologist can also put the nearby hospital on standby (an automated message will be sent). Finally, the cardiologist can send a custom message to the patient, whatever the severity of the situation is, with instructions for immediate actions the patient should take, or for change in medication dosage.

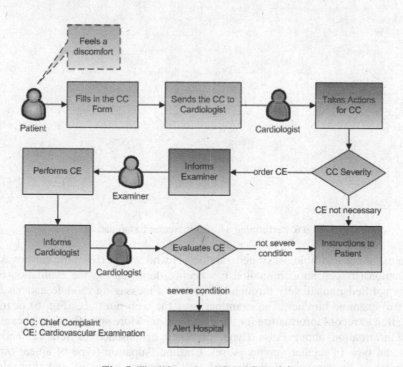

Fig. 5. The life cycle of Chief Complaint

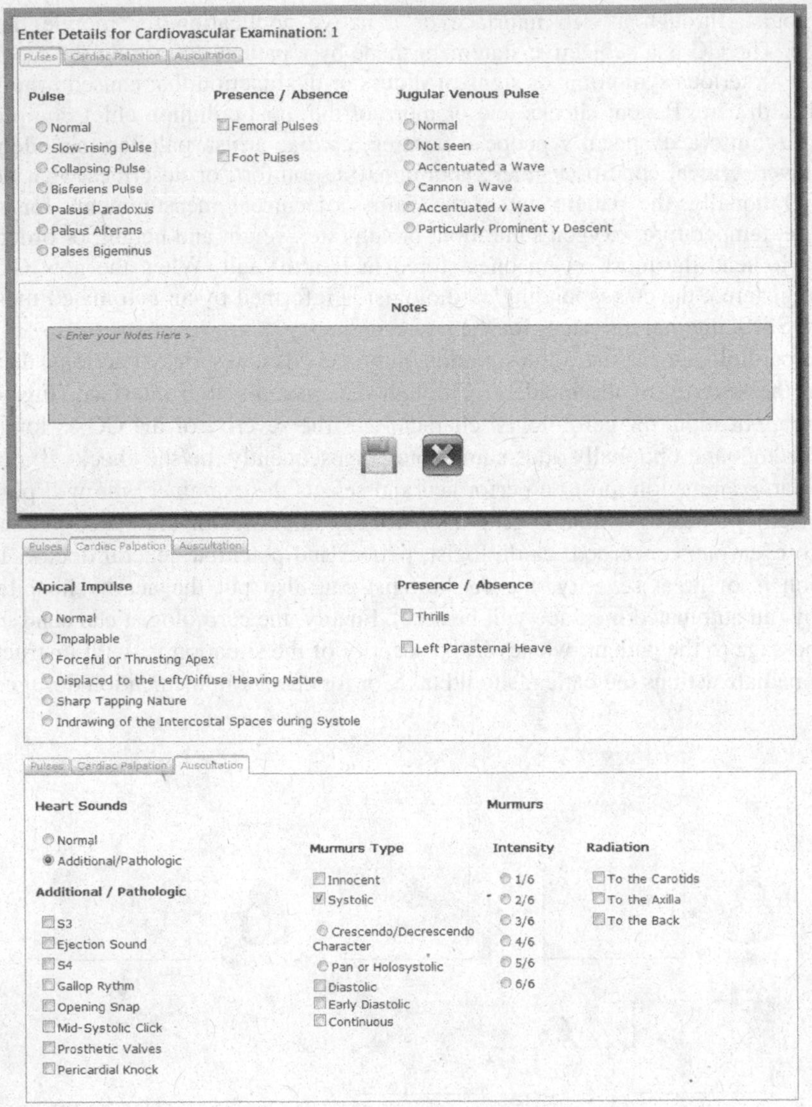

Fig. 6. Performing a Cardiovascular Examination

The examiner (cardiologist or general doctor) who will perform the cardiovascular examination (if such an examination has been ordered during the evaluation of the CC), is notified immediately through the system's messaging module and visits the patient to examine him/her. The examiner uses the web interface (Fig. 6) or mobile application to record information for the examination. More specifically, the examiner enters information about Pulse (type, presence or absence of femoral and foot pulses and type of jugular venous pulse), Cardiac Palpation (type of apical impulse and presence or absence of thrills and left parasternal heave) and Auscultation (type of heart sounds and type, intensity and radiation of murmurs). Through the examination interface, the examiner has the opportunity to view details about the patient's CC. Once the details of the examination have been registered, an automated message is sent to the cardiologist who can directly view all the information and proceed to the necessary actions.

All the aforementioned actions are presented graphically in Fig. 5. This is the life cycle of a Chief Complaint, from the moment the patient feels the discomfort, till the final decision is taken by the cardiologist.

3.3   Measurements of Vital Signs

Depending on the severity of patients' health condition, their cardiologists advise them to perform measurements of their vital signs in regular basis. The measurement types may include blood pressure, blood glucose, oxygen saturation, weight and height and the frequency of measurements depends, in general, on the patient's health condition. The proposed system offers patients tools to record their measurements through web interfaces and mobile applications (Fig. 7). Measurements are performed at home and are imported manually by patients or directly from the medical device (when the device supports connection via Bluetooth or Wi-Fi). Cardiologists have access to patients' measurements for more comprehensive patient monitoring and decision making. Measurements of vital signs are also stored in Microsoft HealthVault.

3.4   Electronic Medical Record

In our system, patients' medical records (EMRs) consist of patient's Medical History, Medication, Laboratory Examinations, Cardiovascular Examinations (physical examinations), periodical Measurements of Vital Signals and Demographics.

The Medical History for cardiology (Fig. 8) consists of detailed information about Coronary Artery Disease (CAD), Intervention for CAD, Hypertension, Heart Failure, Valvular Heart Disease and Heart Rhythm disorders. It also includes basic information about Cholesterol, Diabetes Mellitus, Tobacco use, Family history for heart disease,

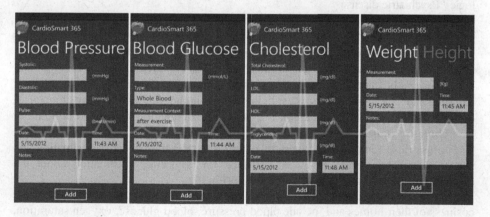

Fig. 7. Mobile interfaces for recording vital signs measurements

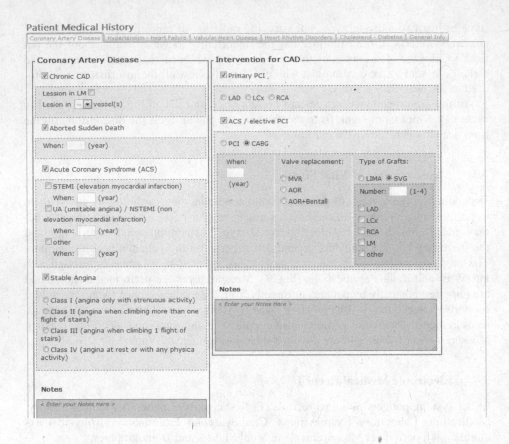

Fig. 8. Recording patient's Medical History

Stroke, Peripheral arterial disease, Thyroid disease, Cancer, Lung / Liver / Kidney / Neurological / Gastrointestinal / Autoimmune / Hematologic / Endocrine / Ophthalmologic / Psychiatric disease.

Medication includes information about patient's current and past medicines, dosage, route, duration, instructions, etc. Medication is updated by patients or cardiologists. This information would also be updated directly by pharmacists, as future potential end users.

Laboratory Examinations include information for complete blood count, coagulation times and biochemical examinations. Information about laboratory examinations is recorded and stored by patients or cardiologists and in the future directly by clinicians.

The Cardiovascular Examination is a physical examination performed at home or in a clinic, by a general doctor or a cardiologist and consists of information about heart pulse (also femoral and foot pulse), jugular venous pulse, cardiac palpation and auscultation. The system offers an interface for recording information about the examination, at the same time it is performed (Fig. 5).

Measurements of vital signs are performed by patients in a regular basis from the cosiness of their homes and include blood pressure, blood glucose, oxygen saturation, weight and height measurements.

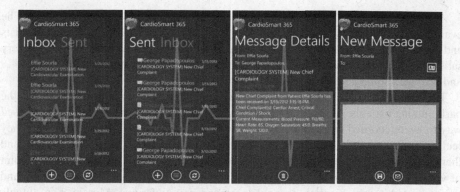

Fig. 9. The mobile interface of the Message Management Module

Demographics include general information of a patient such as name, surname, gender, age, insurance details, contact info and emergency contact details.

Most of the information described above is also stored in Microsoft HealthVault, including Medication, Laboratory Examinations, periodical Measurements and Demographics. This way, the information will be available to third parties after patient's approval. Information characterized as more specific to cardiology, such as the detailed medical history, is stored only in the local database.

3.5   Message Management Module

The proposed system allows interaction and instant communication between the corresponding parties (cardiologic patients, cardiologists, general practitioners) through a message management module. The module manages inbox and sent messages of all end-users and sends automated messages when necessary. The automated messages are formed by customizing specific message templates stored in the system's database. When an automated message is sent or when one user sends a message to another, three actions take place: (a) a message is created internally in the system database and is visible to all recipients; (b) an email is sent to the recipients who have registered their email addresses; and (c) an SMS is sent to the recipients who have registered the numbers of their mobile phones. Through the messaging module interface (Fig. 9), the end user can view his/her inbox, edit and send outbox messages, create, save or send a new message and delete messages.

3.6   System Evaluation

The proposed system is part of collaboration between the Computer Engineering and Informatics Department and School of Medicine of the University of Patras and the General Hospital of Patras "Ag. Andreas" and is currently in testing phase with the following involved parties: a) non-Hospital and Hospital Cardiologists, b) non-Hospital and Hospital General Practitioners and c) Cardiologic Patients.

A first round of evaluation has already taken place during the testing phase. The evaluation is of small scale and refers to the provided services for both cardiologists and patients. The overall satisfaction grade for the services (on a scale of 5 = very satisfied to 1 = very dissatisfied) ranged from 4.0 to 5.0, with an average value of 4.8 and provided some noteworthy results.

The users evaluated the provided services to be very helpful. Most of the participating patients pointed out that the system provides them with a very organised way to keep recorded measurements of their vital signs. Many of them stated that they feel safer and more confident that are closely monitored by their cardiologist and there will be immediate response in cases of emergency. From the cardiologists' point of view, the system provides a complete set of tools for direct patient monitoring and an integrated communication channel in emergency situations. Moreover, a notable proportion of the participants mentioned that it is important to have different access channels (web, mobile) available, by allowing utilization of technology any user can have access to.

4   Benefits

The proposed system was developed to offer advanced added value services for cardiologic patients and physicians into a fast, coherent, flexible and easy to use integrated solution. Its most worth mentioning innovation refers to the tools it provides for early detection of emergency cases and optimal process management of the emergency incident, all in an automated way. The cardiologist is instantly informed about a patient's discomfort and through the system's interface can easily evaluate it and proceed to the necessary actions. In cases of severe conditions where a cardiovascular examination must be performed, the nearby hospital can be alerted in advance, in order to avoid irreversible situations. The message management module plays a supportive but determinant role in this.

Medical Doctors many times need to act under stress and to take decisions in a fast way. Therefore, speed is important for MDs. Our system is a "doctor-friendly" tool that helps recording data in a systematic and quick way, therefore helping MDs to work faster and safer.

Cardiologic Patients can benefit from the system since customized EMRs are at their disposal. Patients can be actively involved in EMRs management since they can update themselves parts of their EMRs (such as medication, laboratory examinations, measurements of vital signs). The benefits of EMRs are widely known, but the customized EMR for CPs contains detailed data, formatted in a way that can be easily understood and quickly viewed even from third party MDs after years. After all, patients suffering from cardiologic diseases, reflect most of the time chronic pathologies that require medication and follow-up for the rest of their life. This way, patients receive lifelong advantageous healthcare.

The system stores and processes a large volume of medical data, making it a useful tool for recording and studying scientifically validated data elements of cardiologic diseases. This data is gathered in one database, continuously updated for scientific research, which is strongly benefited towards better monitoring and understanding of cardiologic diseases.

Last but not least, another field that the proposed system has impact to is the technological one. These advantages derive from the system's architecture design characteristics which are: (a) it is both patient and medical doctor oriented, (b) provides access to end-users through multiple channels (web, mobile), (c) utilizes existing and state-of-the-art medical platforms, such as Microsoft HealthVault, and (d) is easily expandable and deployable.

5   Conclusions and Future Work

In this paper we presented an integrated system for lifelong cardiologic patient monitoring, early detection of emergency cases and optimal process management of the emergency incident. Our system is based on web applications, web services and smart phone applications, as well as on interconnection to Microsoft HealthVault. The system involves the participation of Cardiologic Patients, Cardiologists, General Practitioners, and allows interaction and instant communication between the corresponding parties, as well as Hospitals and out-Hospital health sectors. The benefits of the system concern patients, MDs, everyday clinical practice, Research & Science, and Healthcare Systems.

The proposed solution is part of an ongoing project and it provides full functionality and advanced services to end users for the components described here. However, there is always fertile ground for optimization, since innovative ideas may arise or newly introduced needs have to be met.

Some direct future evolvements include (a) the incorporation of a SMS server component in order to offer an extra channel of communication between end users, as well as an alternative data transfer method between client and server, (b) the integration of a Decision Support module that will incorporate experts' knowledge in order to assist cardiologists in decision making in patient monitoring or when dealing with emergency situations, (c) collection of feedback from end users, in order to incorporate enhancements in future versions, (d) development of a financial tool which by exploiting recorded individual medical costs (from laboratory examinations, medication, physician rewards, patients' transportations, etc) will estimate health costs in an annual basis and make predictions for the forthcoming years. The latter evolvement is of great importance to Healthcare Systems, which are interested in estimating health related financial costs.

Acknowledgements. This research has been co-financed by the European Union (European Social Fund – ESF) and Greek national funds through the Operational Program "Education and Lifelong Learning" of the National Strategic Reference Framework (NSRF) - Research Funding Program: Heracleitus II. Investing in knowledge society through the European Social Fund.

References

1. Boulos, M., Wheeler, S., Tavares, C., Jones, R.: How Smartphones are changing the face of mobile and participatory healthcare: an overview, with example from eCAALYX. In: BioMedical Engineering OnLine, pp. 10–24 (2011), doi:10.1186/1475-925X-10-24

2. Chemlal, S., Colberg, S., Satin-Smith, M., Gyuricsko, E., Hubbard, T., Scerbo, M.W., McKenzie, F.D.: Blood glucose individualized prediction for type 2 diabetes using iPhone application. In: 2011 IEEE 37th Annual Northeast Bioengineering Conference (NEBEC), pp. 1–2 (2011)
3. Gkintzou, V., Papablasopoulou, T., Syrimpeis, V., Sourla, E., Tzimas, G., Tsakalidis, A.: A Web and Smart Phone System for Tibia Open Fractures. In: Cruz-Cunha, M.M., Varajão, J., Powell, P., Martinho, R. (eds.) CENTERIS 2011, Part III. CCIS, vol. 221, pp. 413–422. Springer, Heidelberg (2011)
4. Lindquist, A.M., Johansson, P.E., Peterson, G.I., Saveman, B.I., Nilsson, G.C.: The use of the Personal Digital Assistant (PDA) among personnel and students in health care: a review. J. Med. Internet. Res. 10(4), 31 (2008)
5. Prgomet, M., Georgiou, A., Westbrook, J.I.: The impact of mobile handheld technology on hospital phusicians' work practices and patient care: a systematic review. J. Am. Med. Inform. Assoc. 16(6), 792–801 (2009)
6. Silva, B., Lopes, I., Rodrigues, J., Ray, P.: SapoFitness: A Mobile Health Application for Dietary Evaluation. In: The IEEE 13th International Conference on e-Health Networking, Applications and Services, June 13-15, pp. 375–380 (2011)
7. Sposaro, F., Tyson, G.: iFall: An android application for fall monitoring and response. In: Annual International Conference of the IEEE, Engineering in Medicine and Biology Society, EMBC, September 3-6, pp. 6119–6122 (2009)
8. Sposaro, F., Danielson, J., Tyson, G.: iWander: An Android application for dementia patients. In: Annual International Conference of the IEEE, Engineering in Medicine and Biology Society (EMBC), August 31-September 4, pp. 3875–3878 (2010)
9. Sunyaev, A., Kaletsch, A., Krcmar, H.: Comparative Evaluation of Google Health API vs Microsoft HealthVault API. In: 3rd International Conference on Health Informatics (HEALTHINF 2010), Valencia, Spain, January 20-23, pp. 195–201 (2010)
10. Varlamis, I., Apostolakis, I.: Medical informatics in the Web 2.0 era. In: The 1st International Symposium on Intelligent Interactive Multimedia Systems and Services, Piraeus, Greece, July 9-11 (2008)
11. Hearth360 Cardiovascular Wellness Center (December 2012), https://www.heart360.org/Default.aspx
12. Houston, N.: The Best Medical iPhone Apps for Doctors and Med Students (September 2010), http://blog.softwareadvice.com/articles/medical/the-best-medical-iphone-apps-for-doctors-and-med-students-1100709/
13. Microsoft HealthVault Development Center (May 2012), http://msdn.microsoft.com/en-us/healthvault/default.aspx
14. Newell, D.: 5 Ways Mobile Apps will transform HealthCare (June 2012), http://www.forbes.com/sites/ciocentral/2012/06/04/5-ways-mobile-apps-will-transform-healthcare/
15. QxMD (December 2012), http://www.qxmd.com/specialty/medicine/cardiology-medical-apps-iphone-blackberry-android
16. Webicina - Cardiology in Social Media (December 2012), http://www.webicina.com/cardiology/cardiology-mobile-applications/

Automatic Microcalcification Segmentation Using Rough Entropy and Fuzzy Approach

Burçin Kurt[1,*], Vasif V. Nabiyev[2], and Kemal Turhan[3]

[1] School of Computer Engineering, Department of Medical Informatics,
Karadeniz Technical University, Trabzon, Turkey
burcinnkurt@gmail.com
[2] Department of Computer Engineering, Karadeniz Technical University, Trabzon, Turkey
vasif@ktu.edu.tr
[3] Department of Medical Informatics, Karadeniz Technical University, Trabzon, Turkey
kturhan.tr@gmail.com

Abstract. Microcalcifications have been mainly targeted as the earliest sign of breast cancer, thus their early detection is very important process. Since their size is very small and sometimes hidden by breast tissue, computer-based detection output can assist the radiologist to increase the diagnostic accuracy. This paper presents a research on mammography images using rough entropy and fuzzy approach. Our proposed method includes two main steps; preprocessing and segmentation. In the first step, we have implemented mammography image enhancement using wavelet transform, CLAHE and anisotropic diffusion filter then rough pectoral muscle extraction for false region reduction and better segmentation. In the second step, we have used Rough entropy to define a threshold and then, fuzzy based microcalcification enhancement, after these microcalcifications have been segmented using an iterative detection algorithm. By the combination of these methods, a novel hybrid algorithm has been developed and successful results have been obtained on MIAS database.

Keywords: Image enhancement, rough entropy, fuzzy approach, segmentation.

1   Introduction

Breast cancer is the second most common cause of death among women and remains the only type of cancer with increasing incidence over the last ten years. Furthermore, the longest survival can be achieved if detection is done in early stage [1].

Our study consists two main steps which are preprocessing and segmentation. Preprocessing step includes breast region segmentation, rough pectoral muscle extraction and mammogram enhancement. Segmentation step contains rough entropy based thresholding, fuzzy MC enhancement and then iterative thresholding based Mcs segmentation. We have improved a whole automatic Mc segmentation system which

---

[*] Coressponding author.

M. Bursa, S. Khuri, and M.E. Renda (Eds.): ITBAM 2013, LNCS 8060, pp. 103–105, 2013.

starts from original mammogram image which is more difficult than segmenting Mcs from original Mc ROI image. Our system has been improved using MIAS [2] database and satisfactory results have been obtained.

2   Results

To test our approach on experimental cases we have chosen the MIAS database which consists of 322 mammograms including 208 normal and 114 abnormal images. 28 of 114 abnormal images contain Mcs. Our proposed algorithm does not miss any microcalcification on the MIAS database.

The experimental results of our proposed algorithm can be seen as below:

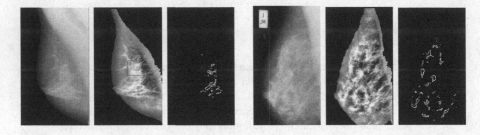

Fig. 1. Original, preprocessed and Mcs segmented images (*mdb213, mdb219 respectively*)

3   Conclusion

As a result a successful system has been developed for Mcs segmentation which does not miss any. In addition to this, a novel hybrid algorithm has been implemented. In the next study, we'll search the false reduction to reduce the Mcs segment results. Furthermore, in the future we'll aim to improve a breast cancer diagnosis system using mammograms.

Acknowledgement. This sudy is a part of a developing breast cancer diagnosis system which is supported as a SANTEZ project by Republic of Turkey Science, Technology and Industry Ministry and AKGÜN Computer Programs and Service Industry Company.

References

1. Guan, Q., Zang, J., Chen, S., Pokropek, A.T.: Automatic Segmentation of Micro-caicification Based on SIFT in Mammograms. In: International Conference on BioMedical Engineering and Informatics, pp. 13–17. IEEE Press, New York (2008)
2. Mammographic Image Analysis Society,
   http://peipa.essex.ac.uk/info/mias.html

3. Kurt, B., Nabiyev, V.V., Turhan, K.: Contrast Enhancement and Breast Segmentation of Mammograms. In: 2nd World Conference on Information Technology (2011)
4. Otsu, N.: A Threshold Selection Method from Gray-level Histograms. IEEE Trans. on Systems, Man & Cybernatics 9, 62–66 (1979)
5. Deepa, S., Bharathi, S.: Efficient ROI Segmentation of Digital Mammogram Images using Otsu's N Thresholding Method. Int. J. of Engineering Research & Technology 2 (2013)
6. Kurt, B., Nabiyev, V.V., Turhan, K.: Medical Images Enhancement by using Anisotropic Filter and CLAHE. In: International Symposium on INnovations in Intelligent Systems and Applications, pp. 1–4. IEEE Press, New York (2012)
7. Bird, R.G., Wallace, T.W., Yankaskas, B.C.: Analysis of Cancers Missed at Screening Mammography. Radiology 184, 613–617 (1992)
8. Kumar, S.V., Lazarus, M.N., Nagaraju, C.: A Novel Method for the Detection of Microcalcifications Based on Multi-scale Morphological Gradient Watershed Segmentation Algorithm. Int. J. of Engineering Science and Technology 2, 2616–2622 (2010)
9. Mohanalin, L., Kalra, P.K., Kumar, N.: An Automatic Method to Enhance Microcalcifications using Normalized Tsallis Entropy. Signal Processing 90, 952–958 (2010)
10. Balakumaran, T., Vennila, I.L.A., Shankar, C.G.: Detection of Microcalcification in Mammograms using Wavelet Transform and Fuzzy Shell Clustering. Int. J. of Computer Science and Information Security 7, 121–125 (2010)
11. Pal, S.K., Shankar, U., Mitra, P.: Granular Computing, Rough Entropy and Object Extraction. Pattern Recognition Letters 26, 2509–2517 (2005)
12. Szczuka, M., Kryszkiewicz, M., Ramanna, S., Jensen, R., Hu, Q. (eds.): RSCTC 2010. LNCS, vol. 6086. Springer, Heidelberg (2010)
13. Mohanalin, J., Beenamol, Kalra, P.K., Kumar, N.: A Novel Automatic Microcalcification Detection Technique using Tsallis Entropy & Type II Fuzzy Index. Computers and Mathematics with Applications 60, 2426–2432 (2010)

A Model for Analyzing the Relation between Potassium (K) and Hemolysis Index (HI) with Clustering Method

Yasemin Zeynep Engin[1,*], Kemal Turhan[1], Sabiha Kamburoğlu[2],
Asım Örem[2], and Burçin Kurt[1]

[1] Department of Medical Informatics, Karadeniz Technical University, Trabzon, Turkey
{ysmnzynp.engin,kturhan.tr,burcinnkurt}@gmail.com
[2] Department of Medical Biochemistry, Karadeniz Technical University, Trabzon, Turkey
yazeyen@gmail.com, aorem64@yahoo.com

Abstract. This study was done to analyze the relation between potassium levels (K) and hemolysis index (HI). A different method for this kind of study -cluster analysis- was used to classify data, according to its hemolysis level. 5 clusters were obtained as a result of the cluster analysis using the absolute differences of first and last K and HI values. Regression analysis was used to understand impact of the hemolysis on K levels at each of 5 different clusters. Results which had HI difference values higher than 295 mg/dL were showed a very high correlation with K.

Keywords: Hemolysis, potassium, regression, clustering.

1   Introduction

Hemolysis is the most common source of preanalytical errors in the clinical laboratories and rejection reason of up to 40-70% test samples [1,2]. Hemolysis may lead to incorrect elevation or decrease in plasma components, such as potassium (K), iron (Fe), magnesium (Mg), lactate dehydrogenase (LDH) and bilirubin [3].

2   Material and Method

In this study, data of patients who treated in various services of Karadeniz Technical University, Faculty of Medicine, Farabi Hospital between 15.06.2011-13.06.2012 was used. Last two potassium (K) and hemolysis index (HI) measurements of blood samples taken from 9777 inpatients were used in this study. Generated data set was subjected to pre-elimination to select the appropriate records. After pre-screening, the number of records decreased to 1252. Selected records were classified by applying cluster analysis, based on the absolute value of differences of the first and last K values and HI values.

---

[*] Coressponding author.

M. Bursa, S. Khuri, and M.E. Renda (Eds.): ITBAM 2013, LNCS 8060, pp. 106–107, 2013.

In this study, absolute values of the differences between last and first K and HI measurements were used as continuous variables for cluster analysis. The distance between clusters was measured in unit of Euclidean using the k-means algorithm. The number of classes was defined 5 for clustering because of HI values classification (HI <= 5 mg / dL 'no hemolysis,' HI <= 30 mg / dL 'ambiguous', HI <= 60 mg / dL 'light', HI <= 200 mg / dL 'medium', HI> 200 mg / dL 'heavy'). Correlation between the average differences of K and HI values in each cluster was calculated.

Table 1. Correlation between the average differences of K and HI values

Clusters	K (mmol/L) (Averages of differences,p=0,00)	HI (mg/dL) (Averages of differences,p=0,00)	n	(%)	r
1	4,925000	1170,250	4	0,31949	0,99
2	3,750000	581,750	8	0,63898	0,80
3	1,102058	29,498	243	19,40895	0,58
4	1,445614	295,772	57	4,55272	0,91
5	0,353936	23,978	940	75,07987	0,54

3   Findings

In 3rd and 5th clusters, mean of differences of HI values were below 30 mg/dL, so that could be said differences of K values were not based on hemolysis. High hemolysis rates were observed in 1st, 2nd and 4th clusters with correlation rates of 0,99, 0,80, 0,91. These clusters are including 5,5% of the data. With regression analysis, in all clusters, a significant relationship was observed ($p<0.01$). A significant and strong positive relationship was seen in 1st ($R^2=0.98$) and 4th ($R^2=0.83$) clusters.

4   Discussion and Conclusion

Clustering provides to take advantage of the existing data more easily. Data can be classified in accordance with the nature of it. This model can be used to calculate a correction factor to estimate K values when it is unable to take blood again.

References

1. Carraro, P., Servidio, P., Plebani, M.: Haemolyzed specimens: a reason for rejection or clinical challenge? Clin. Chem. 46, 306–307 (2000)
2. Lippi, G., Blanckaert, N., Bonini, P., Green, S., Kitchen, S., Palicka, V., Vassault, A.J., Plebani, M.: Haemolysis: an overview of the leading cause of unsuitable specimens in clinical laboratories. Clin. Chem. Lab. Med. 46, 764–772 (2008)
3. Lemery, L.: Oh, No! It's Hemolyzed! What, Why, Who, How? Advance for Medical Laboratory Professionals 15, 24–25 (1998)

A Secure RBAC Mobile Agent Model
for Healthcare Institutions - Preliminary Study

Cátia Santos-Pereira[1], Alexandre B. Augusto[2],
Ricardo Cruz-Correia[1,3], and Manuel E. Correia[2]

[1] Center for Research in Health Technologies and Information Systems - CINTESIS,
Faculty of Medicine of University of Porto (FMUP) - Portugal
[2] Center for Research in Advanced Computing Systems - CRACS,
Department of Computer Science, Faculty of Science of University of Porto - Portugal
[3] Department of Health Information and Decision Sciences - CIDES, FMUP
{catiap,rcorreia}@med.up.pt,
{aaugusto,mcc}@dcc.fc.up.pt

Abstract. Efficient healthcare is thus highly dependent on doctors being provided with access to patients medical information at the right time and place. However it frequently happens that critical pieces of pertinent information end up not being used because they are located in information systems that do not interoperate in a timely manner. There are many reasons that contribute to this grim state of affairs, but what interests us the most is the lack of enforceable security policies for systems interoperability and data exchange and the existence of many heterogeneous legacy systems that are almost impossible to directly include into any reasonable secure interoperable workflow. The objective of this paper is to establish a mobile agent access control model based on RBAC model that allows the exchange of clinical information between different health institutions that fall within the same circle of trust.

Keywords: Mobile agent, RBAC, HIS, Interoperability, Security.

1 Introduction

In this paper we propose a RBAC mobile agent access control model supported by a specially managed public key infrastructure for mobile agents authentication and access control. Our aim is to create the right means for doctors to be provided with timely accurate information, which would be otherwise inaccessible, by the means of strongly authenticated mobile agents capable of securely bridging otherwise isolated institutional eHealth domains and legacy applications [1].

2 Role Based Access Control Model

Our model follows the RBAC structure, where the role keeps the list of possible roles that an agent can assume. The permissions are linked to each different role where its operations are linked into each medical information object.

M. Bursa, S. Khuri, and M.E. Renda (Eds.): ITBAM 2013, LNCS 8060, pp. 108–111, 2013.
© Springer-Verlag Berlin Heidelberg 2013

The standard CEN/ISO 13606-4 [2] defines a set of functional roles and its mapping according to the record component sensitivity. Our proposed model takes advantage of this standard to perform the user-assignment and permission-assignment process. The attributes that mobile agent carries since its creation are used for the external institution agent to attribute a role and assign access permissions.

Furthermore our model includes the break the glass (BTG) mechanism [3] that is an important mechanism to mobile agents when an emergency scenario happens.

3   Mobile Agents: Creation and Migration Process

In order to create the right means to authenticate the mobile agents we had to establish a circle of trust between the health institutions. This circle was formed by the usage of a public key infrastructure (PKI) [4]. In this section we clarified the creation step and how an external should handle with it.

3.1   Mobile Agent Creation Process

The mobile agent creation have to two different steps: (1) the attribute gathering where the necessary attributes (see table 1) are collected in order to proceed the request; and (2) the request validation where the request is cryptographic signed in order to guarantee a non-repudiation between the involved entities.

3.2   Mobile Agent Reception Process

When a mobile agent arrives at an external health institution an external agent receives him by that verifies the mobile agent identity by the usage of the health institution signature attribute. After this process the external agent request the mobile agent cyphered symmetric key attribute in order to obtain the necessary symmetric key to decrypt the common attributes to define which permissions to grant to the mobile agent according to the access control model of the external health institution. Depending on the type of request the external access control could need an approval from an internal member of the institution in order to process the request. In cases like that the external agent provides to the mobile agent an identification number that could be used later to query the status of its requirement. This identification number improves the mobile agent flexibility since the mobile agent could keep his itinerary to other external health institutions and return later to consult the request status.

In special cases where the criticality code is set as emergency the mobile agent will active the BTG mechanism to directly obtain the requested medical information.

4   Case Scenario

To better understand how the agent access control model can be employed in real practice scenario, we exemplified a storyboard to serve as a keystone:

A 32 years old female patient named Inês, from Braga, 38 weeks pregnant, was admitted in the São João Hospital Centre, Emergency Department (ED) with severe abdominal pain. Due to the emergency situation she forgot her pregnancy book at home. Prenatal care was done in Braga Hospital. The doctor who assists the patient in ED, knowing that the prenatal care was done in Braga Hospital triggers an information request to Braga Hospital. He asks for blood analysis, obstetric history, previous pathologies and allergies.

The Figure 1 demonstrates the necessary steps since the agent is creation until the agent return. These steps are described as it follows:

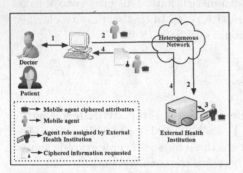

1. Doctor João logs into the HIS that recognizes his role (ED doctor). Then the doctor performs a clinical information request, if doctor João did not have enough permissions, the system would refuse the request. When doctor triggers this request in HIS a mobile agent is created and initiates its migration with a set of attributes presented in Table 1.

Fig. 1. Mobile agent messaging exchange

2. The user (Doctor João) logs into the HIS and the systems recognize his role (ED doctor) Then the doctor performs a clinical information request, if doctor João did not have enough permissions, the system would refuse the request. When doctor triggers this request in HIS a mobile agent is created and initiates its migration with a set of attributes presented in Table 1.
3. Mobile agent arrives to the external institution (Braga Hospital). The external agent authenticates the mobile agent by verifying the signatures attributes to ensure that its legitimate.
4. After perform authentication, the external institution RBAC module assign a role with access permissions. Since Doctor João is an emergency doctor in São João Hospital Centre and the reason appointed is care provision, the mobile agent will assume the Privileged healthcare professional role that can access almost all the patient information like demonstrated in Table 1. Since the authorization process succeeded, the mobile agent receives an authorization token to submit its query to the external agent.
5. Once finished, the mobile agent receives the results of the query and departs from external institution back to its home institution.

Table 1. Example of mobile agent non-ciphered attributes

Attribute	Value
User Id	43259823PRT
User role permission	ED doctor
Data query	([Blood analysis, obstetric history and allergies])
Patient id	PRT12343652
Criticality code	1
Time to response	7200000 milliseconds (2hours)
Reason code	01 (care provision)
List of external institution	([network host address, Braga Hospital certificate])
Description	38 weeks pregnant, admitted in São João Hospital Centre ED due to abdominal pain. Lacks pregnancy book.
Requester signature	ASd2qFHDFGg3g43g46G323sEa...We3
Health Institution signature	Juy7jgjT6rhgtg5SDFe3egt34FRd...DYJ

5   Conclusion

The consequence of unauthorized disclosure of health-related information may fatally affect a patients health, employment prospects and social standing. The main contribution of this work was to guarantee a secure communication channel between health institutions by the means of an access control for mobile agents.

This work is an initial proposal, the next steps are implementation and evaluation of our proposed model within a specific case study in a real healthcare institution, more precisely on São João Hospital Centre, which is the second biggest hospital in Portugal.

Acknowledgments. This work was financed through the project SAHIB [PTDC/EIA-EIA/105352/2008].

References

1. Vieira-Marques, P.M., Cruz-Correia, R.J., Robles, S., Cucurull, J., Navarro, G., Marti, R.: Secure integration of distributed medical data using mobile agents. IEEE Intelligent Systems 21(6), 47–54 (2006)
2. CEN/ISO 13606-4. Health informatics - electronic health record communication (2009)
3. Ferreira, A., Chadwick, D., Zao, G., Farinha, P., Correia, R., Chilro, R., Antunes, L.: How securely break into rbac: the btg-rbac model. In: Proceedings from 25th Annual Computer Security Applications Conference, ACSAC 2009 (2009)
4. Santos-Pereira, C., Augusto, A.B., Correia, M.E., Ferreira, A., Cruz-Correia, R.: A mobile based authorization mechanism for patient managed role based access control. In: Böhm, C., Khuri, S., Lhotská, L., Renda, M.E. (eds.) ITBAM 2012. LNCS, vol. 7451, pp. 54–68. Springer, Heidelberg (2012)

Adaptive Model of Cardiovascular System: Realization and Signal Database

Jan Havlík[1], Miroslav Ložek[1], Matouš Pokorný[1],
Jakub Parák[1], Petr Huňka[2], and Lenka Lhotská[2]

[1] Department of Circuit Theory, Faculty of Electrical Engineering,
Czech Technical University in Prague, Technická 6, CZ-16627 Prague 6
[2] Department of Cybernetics, Faculty of Electrical Engineering,
Czech Technical University in Prague, Technická 6, CZ-16627 Prague 6
xhavlikj@fel.cvut.cz

Abstract. The paper deals with a study of relationship between hemodynamic parameters and other various vital signs based on the modelling of hemodynamic parameters. The design of an adaptive mechanical
model of cardiovascular system is presented in the paper. The connection between the modelling of cardiovascular system and smart homes
and ambient assisted living applications is also discussed.

Keywords: cardiovascular system, vital signs, telemetry.

1 Introduction

The poster deals with a study of relationship between hemodynamic parameters
and other various vital signs based on the modelling of hemodynamic parameters.

Modelling is one of the approaches to understanding complex biological
systems. In the field of biomedical engineering it is important to have robust
background knowledge about dependencies between various vital signs. As cardiovascular diseases are very frequent with the elderly in developed countries,
the modeling of cardiovascular system and the connection between hemodynamic
parameters and other vital signs seems to be a very important task.

2 Methods and Results

The system we introduce in the work is designed as a combination of a telemetry system and a mechanical model of cardiovascular system. The main task
of the telemetry system is to monitor vital signs, e.g. electrocardiogram (ECG)
and peripheral photoplethysmogram (PPG). The other considered parameters
are the phonocardiogram (FCG), breathing activity, transthoracic bioimpedance
and physical activity sensed by accelerometers. The system consists of electrodes
and sensors for monitoring of vital signs, input modules converting the sensed
physical values to the electrical values, microprocessor unit and wireless communication module (bluetooth or wi-fi depending on required distance between

M. Bursa, S. Khuri, and M.E. Renda (Eds.): ITBAM 2013, LNCS 8060, pp. 112–113, 2013.
© Springer-Verlag Berlin Heidelberg 2013

sensed person and controlled model). The model of the cardiovascular system is a set of pipes representing arteries and veins, the pump representing the heart, sensors and the electronic circuits serve as hemodynamic parameters monitor and pump driver. The model is able to change of the system parameters such as heart rate, cardiac output and resistance of the vascular system. The system allows measuring various hemodynamic parameters e.g. blood pressure, blood flow, pulse wave velocity, type of the flow etc.

The poster presents the realization of the system and the signals acquired with the model.

3   Conclusion

The complex system connecting the telemetry sensing of vital signs and the mechanical modelling of the cardiovascular system has been designed and being realized. The system increases the possibilities to study relationships between vital parameters and to improve existing and develop new methods for classifying life threatening situations. These methods could be effectively used in the smart homes facilities and in the ambient assisted living.

Acknowledgement. This work has been supported by the grant No. SGS11 /153/OHK3/3T/13 of the Czech Technical University in Prague, grant No. SGS13/203/OHK3/3T/13 of the Czech Technical University in Prague and also by the grant No. G3 902/2013 presented by the University Development Foundation.

References

1. Batzel, J.: Cardiovascular and respiratory systems: modeling, analysis, and control. Society for Industrial and Applied Mathematics, Philadelphia (2007)
2. Cobelli, C., Carson, E.: Introduction to modeling in physiology and medicine. Academic Press (2008)
3. Darowski, M., Ferrari, G.: Comprehensive models of cardiovascular and respiratory systems: their mechanical support and interactions. Nova Science, New York (2010)
4. Westerhof, N., Stergiopulos, N., Noble, M.I.M.: Snapshots of hemodynamics. Basic Science for the Cardiologist (18), 121–126 (2005)

Author Index